HOW TO BUILD *Max-Performance*
MITSUBISHI 4G63t
Engines

Robert Bowen with Robert Garcia of Road/Race Engineering

CarTech®

CarTech®

CarTech,® Inc.
39966 Grand Avenue
North Branch, MN 55056
Phone: 651-277-1200 or 800-551-4754
Fax: 651-277-1203
www.cartechbooks.com

© 2009 by Robert Bowen

All rights reserved. No part of this publication may be reproduced or utilized in any form or by any means, electronic or mechanical, including photocopying, recording, or by any information storage and retrieval system, without prior permission from the Author. All text, photographs, and artwork are the property of the Author unless otherwise noted or credited.

The information in this work is true and complete to the best of our knowledge. However, all information is presented without any guarantee on the part of the Author or Publisher, who also disclaim any liability incurred in connection with the use of the information.

All trademarks, trade names, model names and numbers, and other product designations referred to herein are the property of their respective owners and are used solely for identification purposes. This work is a publication of CarTech, Inc., and has not been licensed, approved, sponsored, or endorsed by any other person or entity.

Edit by Travis Thompson
Layout by Christopher Fayers

ISBN-13 978-1-61325-066-2
Item No. SA148P

Library of Congress Cataloging-in-Publication Data

Bowen, Robert
  How to build max-performance Mitsubishi 4G63t engines / by Robert Bowen.
    p. cm.
  ISBN 978-1-932494-62-4
  1. Mitsubishi automobiles--Motors--Modification. 2. Mitsubishi automobiles--Performance. I. Title.

TL215.M54B69 2008
629.25'04--dc22

                              2008032649

Printed in USA

Back cover top left: This 4G63t is in a Galant VR-4, one of the first 4WD super-sedans available in the USA. It's a great car, and with a few well-chosen modifications, it can hold its own even against the latest Evo IX and competitors.

Back cover top right: A few thoughtfully chosen modifications to the outside of your engine can make your car faster and easier to tune later without taking away any reliability. Choose good-quality parts from reputable manufacturers and install them with care and your engine will look like this, and perform better.

Back cover middle left: Any of the 16G turbos makes an excellent street turbo, which makes sense since they were all originally used on stock 4G63t applications in other markets. The largest of the family, the famous Evo-III 16G shown here on the left, is a 350+ hp turbo.

Back cover middle right: The various 1g ECUs are similar, but not all of them have an EPROM. You will have to pull the lid off and look in the bottom right corner to find out if yours can be tuned. This one has been "chipped" with a modified chip programmed for a particular combination of parts. Swapping chips is not as convenient as reprogramming a DSMLink adapter from a laptop, but it works just as well.

Back cover bottom left: The 4G63t has a nice combustion chamber by modern standards, compact with little surface area and a dished piston. The four valves mean that valve shrouding is not really a problem.

Back cover bottom right: The 4G63t's block is one of the engine's great strengths. It is rather low-tech, has lots of material in all the right places, and is nearly unbreakable absent abuse. Subaru, Honda, and Nissan tuners can only dream of the power levels achievable with a mildly built 4G63t short block and the right supporting parts.

TABLE OF CONTENTS

AUTHOR BIOS

Robert Bowen is a freelance writer and an automotive product specialist for an advertising agency in Los Angeles. He is a former import car technician who decided writing about, and studying, cars was more lucrative than working on them. He is the author of *How to Rebuild and Modify Manual Transmissions*, numerous articles for *Grassroots Motorsports* magazine, and other publications in the United States and United Kingdom. He is a member of the Motor Press Guild. He's also been an occasional competitor in autocross, track events, and drag races, and a builder and restorer of dozens of cars and motorcycles. When he isn't behind a computer screen, he can be found photographing cars and events in Southern California for *Grassroots Motorsports* or other automotive clients.

The co-author of this book, Robert Garcia, is the shop manager and engine builder at Road/Race

Robert Bowen

Robert Garcia

Engineering (RRE) in Santa Fe Springs, California. RRE is one of the best-known Mitsubishi turbo tuning shops in the United States, with experience tuning the 4G63t to com-

pete in nearly every form of motorsports. Robert Garcia has built thousands of 4G63t engines over the years, from the early DSM days through the Evo VIII and IX.

FOREWORD

BY ROBERT GARCIA OF ROAD/RACE ENGINEERING

My years of building 4G63t motors has been a great experience. I've learned a lot and have been able to do things and go places that I never expected. I started out working with Mike Welch of Road/Race Engineering out of his garage. A lot of great opportunities came up along the way, like attending road races and rallies all over the country and some as far away as Mexico. The best thing that I've learned though is how to figure out things for myself. Someone can tell you how to do something but you'll remember it much better if you figure it out yourself, no matter how frustrating that is. Learn by doing—that's what got me started in this business.

Mike Welch and I started out doing preparation and servicing on rally cars. We both had a background in bodywork, and since fixing crashed bodywork leads to fixing mechanical stuff too. We eventually learned a lot about the cars we were working on—Mitsubishi AWD turbo cars and the Mazda 323 GTX. In the beginning we didn't know too much about engine building, but we learned fast. There weren't many other people who knew these cars

back then either, so we got a lot of experience the hard way.

Things really changed for us when we bought all the cars and parts from team Mitsubishi. We got these cars—four, I think—and ten truckloads of parts. We started trying to put the cars back together and make them run, so that we could use them, sell them, or race them. That was the best opportunity we had to really learn about them. There were enough extra parts around that we could try different things, and if they didn't work out we would have enough parts to try it again. Eventually we put a couple of cars together for us to race, or have other people race. We did some road racing but mostly ran rally events, and the cars kept running.

Pretty soon, word got out that we had all these parts and knew the cars. Since it's the nature of turbo cars to blow up when people drive them hard, people started showing up with blown up motors and transmissions.

They would ask us, "Can you fix this?" Well, we figured we've done it once, we can do it again. After that, more people started coming to

us with their Mitsubishi turbo cars. It was the mid 1990s, nobody local really knew how to work on them back then, so we got a lot of cars with timing belts done wrong, and we figured out how to do it right.

The first motor I built was for our Eclipse race car that was going to race against a bunch of Porsches and other sports cars at the Grand Prix of Los Angeles—the downtown L.A. street race. The car actually did good; It finished 6th place out of a 20-car field of much more exotic cars, and so I just kept building motors for people. That lasted through the first shop we had in Los Alamitos, through our shop in Huntington Beach, and to where we are now.

After all these years I still enjoy building motors. Sometimes I build transmissions just to do something different, but I like building motors because I learn something new every time I do it. No two cranks are the same—some have tight thrust bearings, some have tight endplay on the rods, some don't. There are different ways to approach each build—sometimes swapping parts around helps and sometimes it doesn't. When you start combining parts from different manufacturers you can end up with different clearances.

One thing I've learned is to always start with a clear head. It's very easy to overlook something when you are stressed, tired, hungry, or angry. Take your time and double check everything. Try to do it right the first time because it's cheaper and easier to do it once. It's never any fun to pull a motor out of a car and re-do something because of a careless mistake. Paint marks on torqued nuts and bolts save time and ensure nothing is gonna fall off.

I've been lucky since I've been able to try out a lot of different parts over the years. Sometimes people bring in parts from every corner of the Internet and ask us to build their motor. I have my ideas about what works, but it's always interesting to get to test different parts on the customer's dime. Of course I prefer to use what I have experience with. There is a reason we sell what we sell at Road Race—we know it works.

When I have a choice, I tell people not to skimp on external parts like water pumps and timing belts. OEM parts have proven to be the most reliable in that area. Aftermarket rods and pistons are great, but I recommend using Mitsubishi gaskets and everything else because it works. There's an aftermarket manufacturer for everything, and of course people sell what they have or what they make the most money on, not what's best for you or your project.

The most important part of any engine build is good machine work. You've got to have a properly surfaced block and head or you will be taking the engine apart and doing it again. In testing I have seen how much the cylinder walls move with different types of head gaskets under torque. You must bore the 4G63t

block using a torque plate and the same type of head gasket that you will be using in the buildup. A proper line bore is also critical to good main bearing clearances. Also the timing belt has to be set up right—get the shop manual and read it while you build your engine. Vfaq.com has an excellent write up on that also.

If you're starting from scratch, the 2.0-liter blocks seem a lot better than the 2.4-liter blocks. I don't consider the 2.4 the hottest ticket. The block just isn't suited for high-horsepower applications; for motors heading for over 400 hp keep the 2.0 block. Structurally the 2.0 is a much better design. A lot of learning on that was on my dime and other people's dime. The long-stroke 2.3 seems to work best— it's a really good combination.

Of all the stock motors, I like building EVO motors because the parts are the most developed, like the 4-piece thrust bearing, and the resulting motor is a lot stronger. I enjoy running really high-horsepower motors that I have built on our Dynapack chassis dyno because I can get feedback quickly about what works and what does not.

As the EVO X becomes more popular I look forward to building the 4B11 motor—it's very different in so many ways. The timing chain is nice and it should make for a more robust motor, since I've seen a lot of timing belt failures over the years. But I don't see the 4G63t motors going away anytime soon. They make lots of horsepower and there are tens of thousands of these cars still around.

That's where this book comes in— with this hopefully more people will build these motors, or at least understand how they're put together. I had always thought it would be good to write a book about 4G63t motors. I know it would have been great to have a book like this in the beginning.

When Rob Bowen approached me and asked if I wanted to work with him on this book, it seemed like the right thing to do. Sometimes the process of working on this book has been stressful, but I've enjoyed it. It's just another one of those experiences I never expected to have. Thanks, Rob, for getting me involved. It was a pleasure to work with you, and I got a new motorcycle buddy out of the deal.

Big thank you to Dave Wolin for introducing us to the world of Mitsubishis, and to Mike Welch for sharing all his knowledge and his never ending support over many years, in the rollercoaster of working in this business. Justin Dubois For his pioneering efforts on the 1g in a 2g swap. And to my crew: Amphol Tongkul AKA "Lod Doggie," Chris Encinas, Akaruk Thavilyati AKA "Art," Rueben Olague, and Sam Chaysavang (the new guy). I couldn't ask for a better team to work with.

And, to the only three guys that would agree to sit in the passenger seat of my rally car and co-drive: Robert Arriaga, Tony Vu, and Ama Sehmi, you guys are nuts. Thanks for all your efforts. To all the guys who came out to the rallies and serviced with us: Jeremy Siglar, Joey Siglar, Dave Kwee, Saeed Ettefogh (I hope I didn't leave anyone out). Thanks for all the support. We had lots of great times out in the dirt. And Jacquie Wagner: Thanks for all your help since day one. We couldn't have done it without you.

Last but not Least. I dedicate my part of this book to my Family: my Grandma Mona, my mom and pop, Donie, the love of my life Gina, Keena, Iree Marie, Dusty, and Dean. Thank you all for your support.

— Robert Garcia, Road/Race Engineering

ACKNOWLEDGMENTS

This project would never have happened without the help of many people, including friends, family, and industry contacts who have provided me with support, help, and information over the past several years. There's no way to thank all of you, so I hope that you will forgive me if I've omitted you from this list.

First, my wife, Nichole, has always been as supportive as possible. She's always inspired me to keep going even when I'm ready to throw in the towel.

Robert Garcia has been crucial to this book, sharing a small portion of his incredible knowledge of these engines with me. Without him, this project would never have gotten off the ground. Thanks also to Mike Welch for opening up his shop and letting me pester his employees with question after question. Robert Ramirez, aka "Honda Robert," has also been incredibly helpful. I'm pretty sure I didn't get any special treatment from him— he answers more questions about these cars on a daily basis than anyone I know.

Tom Dorris helped by providing information on his excellent and innovative DSMLink product, and by helping to fact-check the ECU and Fuel System chapters. Meeting Garrett Engineer Willi Smith was a one-in-a-million chance, and his work reviewing my turbo chapter made sure I wasn't skimping on the science.

Mitsubishi Motors North America was helpful during my research as well. Thanks to Moe Durant and Mike Evanoff for their tireless support of the Lancer Evolution. I'd also like to thank Mitsubishi Motors Corporation in Japan; without them, there wouldn't be the 4G63t to write about!

The staff at CarTech, including Peter, Josh, and the rest, has been very patient with me through numerous changes, revisions, and delays. Also, as always, my editor at *Grassroots Motorsports*, David Wallens, deserves plenty of thanks for giving me my first experiences in this industry. Thanks to all of the other media outlets that have given me the opportunity to share some of my experience with readers.

INTRODUCTION

A book about building one of the most popular and well-developed high-performance engines of the last 20 years is a tough thing to write. The 4G63t has been with us in various forms long enough to have been tweaked and tuned by just about everybody. The basic strengths and weaknesses have been found and exploited many times. People who are a lot smarter and more experienced than I have spent thousands of hours tweaking and racing this powerplant, and writing about their experiences.

That does not, or should not, mean that there is nothing more to discover or say about the 4G63t engine. Although this engine is no longer manufactured by Mitsubishi, thousands and thousands of these motors were built over the years and will be powering our street, drag, rally, and road-racing cars for many more years to come.

Because there is so much to know and learn about this engine, I partnered up with Robert Garcia of Road/Race Engineering in Santa Fe Springs, California. RRE is the most well-known and respected Mitsubishi-turbo performance shop in the United States, and Garcia is their in-house engine builder. He's been building turbocharged Mitsubishi engines for more than 10 years, starting with preparing the first DSMs for rally racing. At the time, Robert teamed up with Mike Welch and Robert Tallini of Road/Race Engineering to bring his skills to the masses.

After others started to notice how fast and reliable his motors were, they started asking Garcia to build engines for them. Working out of RRE, he developed a far-ranging reputation as a 4G63t expert, especially for his pioneering work with 2.3- and 2.4-liter stroker motors.

Robert has agreed to share a sliver of his experience with me, and with the readers of our book. No, you won't be able to build engines like he does, but you will definitely get an idea of what it takes to build maximum-performance 4G63t engines. This book, though it is one among many performance books out there, will hopefully fall into the category of indispensable for most people. The information here can certainly be found in other places, but you will have to look long and hard to find a take on building

Mitsubishi turbo engines that is as readable and useable.

There is a reason the title is *How to Build Max-Performance Mitsubishi 4G63t Engines*. The primary subject matter for this book is the Mitsubishi 4-cylinder turbo engines from the late 1980s to the present. These tough and powerful engines were found in sedans and sports cars including the Galant, Eclipse, Laser, Talon, and EVO, among others. This does not mean that the book will be useful only to people with these kinds of cars. Modern overhead cam engines are surprisingly similar, so the same tricks that work on one will work on most.

Also, this book is also not a manual for rebuilding your stock engine. For that, you'd be better off finding a copy of the factory shop manual for your Mitsubishi. Rather, this book is the perfect companion for that shop manual, especially if you intend to increase the performance of your engine while you rebuild it. Read this book from cover to cover before you

start on your engine project, and keep it handy while you work.

This book will also be helpful as an introduction to high-performance 4G63t engines, for those who plan to pay someone to build an engine for them. After reading this book, you should be able to tell your engine builder exactly what you want from your engine, and know exactly what to expect. It will also help you talk to machinists, head porters, cam grinders, and engine management tuners in their respective languages.

There is no way to anticipate the level of knowledge and experience of every reader, but this book was written to appeal to every level, from beginning tuner to experienced pro. The most technical sections have been presented in an easy-to-understand format, but the information is useful for anyone dealing with engine technology. It is hoped that every reader will take something away from each chapter, whether the material is directly applicable to your situation or not.

INTRODUCTION TO THE 4G63t

The 4G63t is a member of Mitsubishi's Sirius engine family. For a time, Mitsubishi named its engine families after distant stars such as Astron, Orion, and Saturn. Sirius engines follow the standard Mitsubishi engine codes: the first character (4) refers to the number of cylinders. The second (G) refers to the fuel, in this case gasoline. The third character refers to the engine family, and the forth refers to the engine itself; each engine is given a distinct number that is unrelated to engine capacity. The final character (t) is optional, and indicates that the engine is turbocharged.

Sirius Engines

Most of the Sirius engines are equipped with balance shafts to smooth the secondary vibrations inherent in a 4-cylinder engine. They are all overhead-cam designs, with belt-driven cams and water pump. Sirius engines share the same head bolt pattern and one of two bell-housing bolt patterns, though motor mount bosses and block castings vary a little between engines. Some are turbocharged, but most are not.

The 1974 Lancer GSR 1600 was one of the first high-performance Mitsubishis. By today's standards it was painfully slow, but the 1,600-cc 4G32 engine was powerful and reliable; reliable enough to give it several rally wins in the mid 1970s.

The cylinder heads of each Sirius engine vary the most, despite having the same bore centers and head bolt pattern. From the earliest single-cam motors to the latest Lancer Evolution, there is almost no family resemblance at first glance. Heads came with two or four valves per cylinder, and single or double overhead cams. Some internal parts are interchangeable between the various Sirius engines. The specifics are detailed in later chapters.

4G61

In 1990, Mitsubishi began selling the smallest Sirius engine, the 1.6-liter (1,595 cc) 4G61, in the USA. Turbo and non-turbo variants powered the 1990 to 1992 Mirage turbo

---

and Mirage. The 4G61 was a dual-cam engine with 16 valves, a bore of 82.3 mm, and a stroke of 75 mm. Neither version had a balance shaft, as the pistons and rotating assembly were so lightweight it was deemed that one was not necessary.

A relative of the 4G61 non-turbo engine was also used in the 1992–1995 Hyundai Elantra. The engine is not particularly high-powered, but it shares a bellhousing bolt pattern and many mounting points with the 4G63t engine. This makes the lightweight Hyundai and Mirage cars perfect donors for a stealthy road rocket.

4G62

The 1.8-liter (1,795 cc) 4G62, one of the oldest members of the Sirius family, was introduced to the USA market in the early 1980s. It arrived under the hood of the rare (in the USA) Cordia turbo, produced from 1984 to 1989. It displaced 1.8 liters with an 80.6-mm bore and an 88-mm stroke. The turbocharged variant produced somewhere in the range of 140 hp. The head was a single-cam

The Mitsubishi Starion and its twin, the Chrysler Conquest, were the company's first foray into turbocharged performance in the US market. These were fast, good-handling cars that got enthusiasts thinking "Mitsubishi" and "performance" in the same sentence.

design, and was installed in a front-wheel-drive chassis.

4G63

The first 2.0-liter (1995 cc) 4G63

The Starion's front-engine, rear-drive platform was more common in the mid 1980s than it is now. The turbo 4G54 engine made good power but some parts of the powertrain were less than optimal. Many of these cars survive with 4G63t swaps using a variety of donor parts.

was introduced at the same time as the 4G62 in the decidedly uninspiring and non-sporty Dodge Colt Vista wagon. It was a single-cam motor with a carburetor and no turbo. It was not until the introduction of the Eclipse, Eagle Talon, and Plymouth Laser in 1989 that the 4G63 sprouted both a turbocharger and a second camshaft for the USA market, though Japanese-market versions existed a few years earlier. All variants of the 4G63 have an 85-mm bore and 88-mm stroke.

4G64

From 1988 on, naturally aspirated Galants and a few other Mitsubishi models were equipped with one of the largest Sirius engines: the 4G64. Its 86.5-mm bore and 100-mm stroke gave it a nearly 2,350-cc in displacement. There were several cylinder heads installed on this engine starting with a SOHC 8-valve head, and moving on to various 16-valve

SOHC and DOHC heads. None were turbocharged. In some non-USA models, the 4G64 had GDI, a fueling system that injects gasoline directly into the combustion chamber for improved fuel economy and midrange torque.

4G67

A nearly forgotten member of the Sirius engine family, the 4G67 was a 1.8-liter (1,836-cc) DOHC engine used by Hyundai in the mid 1990s Elantra. No USA-market Mitsubishis used the 4G67 engine, although it was used to power some European-market Mirage/Lancer models. The bore is 81.5 mm, with an 88 mm stroke.

4G69

The newest member of the Sirius family is the 4G69 engine, a 2,378-cc version with an 87-mm bore and 100-mm stroke. The 4G69 head has dual overhead cams, roller rockers, and Mitsubishi's MIVEC variable valve timing system (similar to the final 4G63). While it shares many parts with the 4G64 and 4G63, it's different enough that parts can't be swapped easily between them.

Until the recent (2007) advent of the Mitsubishi "world" engines, developed with Chrysler and Hyundai, the 4G69 was the backbone of the USA market. From 2004 and up, Eclipse, Outlander, Lancer, and Galant models were all equipped with this engine.

The 4G63t in the USA

As mentioned above, all variants of the 4G63 engine have an 85-mm bore and 88-mm stroke, but that is about all they have in common. The earliest 4G63 engines had just one camshaft, a distributor, and a carburetor. The final ones had two cams, variable valve timing, a huge tur-

bocharger, and fuel injection. In total, there are at least two bellhousing bolt patterns, half a dozen heads and pistons, two kinds of crankshafts, different main bearing caps, and numerous detail differences. The turbocharged 4G63t engine adds more variation—there are several heads, turbochargers, intake and exhaust manifolds, and other parts. For the determined parts swapper, the 4G63 and 4G64 family can provide an almost unlimited number of combinations.

The first 4G63t in the USA market came about as a direct result of Mitsubishi's participation in rally racing. In the late 1980s, Mitsubishi wanted to compete in the FIA Group A rally circuit. Rules at the time required the use of a production-based, 2.0-liter, turbocharged powertrain driving all four wheels. To qualify as production-based, some 5,000 units were required to be built, so Mitsubishi designed a special rally homologation model of their midsize Galant sedan.

It was clear that the Galant engine, the 4G63t, was designed from the start as a race-ready motor. It had beefy forged connecting rods,

The production Galant VR-4 was introduced to Japan in 1988 with the first turbocharged, roughly 200-hp variant of the 4G63. Total production soon exceeded the 5,000 units required as the Galant's rally wins piled up and Mitsubishi exports grew.

dual overhead camshafts, four valves per cylinder, and a well-sized turbo for a street car. Mounted transversely on the left side of the car, it drove a transmission on the right side, and had six bolts holding the flywheel to the crankshaft.

The USA market did not immediately receive the Galant-based hot rod. Instead, the first 4G63t showed up in the new Mitsubishi Eclipse, Eagle Talon, and Plymouth Laser sports coupes. Diamond Star Motors was the

The USA-market Galant VR-4 was an amazing super-sedan, one of the first to boast four-wheel drive and a turbocharged engine. Even today it makes an interesting and unique performance car, with fewer than 3,000 imported over the car's three-year run in the USA.

The first USA-market use of the 4G63t engine was in the 1989 Mitsubishi Eclipse and its platform-mates, the Eagle Talon and Plymouth Laser. All three were well liked when they were introduced, particularly in the areas of technology and looks.

name of the Mitsubishi-Chrysler joint venture that brought the triplets to the USA market in 1989, and these early cars are often referred to as "DSMs" for the initials of their manufacturer. In this text, we will sometimes just use "Eclipse," but rest assured we also mean Talons and Lasers.

The engine installed in these first-generation cars (usually abbreviated as "1g") was rated at 195 hp. The limited edition Galant VR-4 finally made it to the USA market from 1991 to 1993, and carried a 4G63t similar to the one in the 1g Eclipse.

In April 1992, the 4G63t was heavily revised to match the engine used in the Japanese-market Lancer Evolution. Different oil sprayers were added to cool the piston crowns, the pistons and rods were lightened, and the crankshaft used a different, seven-bolt attachment to the flywheel, but it was otherwise similar to the earlier engine.

In 1995, the Laser disappeared, and the Eclipse and Talon received a major makeover. Included in the second-generation (or "2g") facelift was a major engine update. The turbo,

intake runners, and throttle body all shrank in size, but output increased by 15 hp. Now rated at 210 hp, the engine gained tractability and midrange torque. The smoothed-out plumbing and smaller turbo helped the car's drivability compared to the 1g. The power increase was due to higher boost. The engine was largely unchanged for the rest of the 2g DSM production.

Unfortunately, in 2000 Mitsubishi introduced the third-generation Eclipse without a turbo variant. With Diamond-Star Motors now defunct, the 1999 Eclipse was the last turbocharged Mitsubishi for nearly five years. The naturally aspirated members of the family, the 4G63, 4G64, and 4G69 engines, continued to power many of Mitsubishi's North American models in the meantime, but none had the same punch or were as easy to upgrade as the turbo motor.

Even after Mitsubishi North America discontinued the engine, the 4G63t quickly developed a reputation for being tough and powerful. Hundreds found their way to the race track in one form or another, usually in rally or road racing. During the 1990s, the small import drag racing scene on both coasts discovered the turbo Mitsubishi and its siblings and the aftermarket for the 4G63t took off.

The Lancer Evolution

While the 4G63t and its AWD drivetrain was turning USA-based tuners and racers into fans, Mitsubishi continued to develop it as a rally racing powerplant. When the Galant VR-4 was dropped in 1993, the 4G63t and its racing development were transferred directly to the subcompact Lancer sedan. Sold as the Mirage in the USA, the Lancer was an unremarkable naturally aspirated grocery getter. But when fitted with the

mighty 200-hp 4G63t and four driven wheels, it turned into a rally monster.

The newest rally homologation special, the Lancer Evolution, became the Mitsubishi performance car in international and domestic markets. It was a successful competitor in world rally racing, but was never imported to the USA. Although USA enthusiasts clamored for more lightweight, fast cars with four wheel drive and decent handling, Mitsubishi did not listen, even after it dropped the Eclipse turbo.

The first Lancer Evo, as it was soon nicknamed, the Evolution I, was basically the same drivetrain as the Galant VR-4 squeezed into the Lancer floor pan. The engine was tweaked to produce 244 hp and 228 ft-lbs of torque with a different turbo and exhaust manifold. The Evo I was sold between 1992 and 1993, and was intended solely for the Japanese market. Only 5,000 examples of each of the first three Lancer Evolution models were produced to adhere to FIA Group A regulations.

In 1993, Mitsubishi introduced the second Evo, the Evo II. It was based on the same Lancer platform as the Evo I, but the engine, drivetrain, bodywork, and suspension were

The Lancer Evo drove the development of the 4G63t powerplant throughout the 1990s and turn of the century. With generations numbered sequentially in roman numerals from I to X, the Evo became one of the most-recognized and celebrated rally cars in the world. (Image courtesy MMNA)

further tweaked. Horsepower grew to 252, while torque was unchanged. The 4G63t that powered both Evolution I and II models was very similar to the 2g Eclipse motor sold in the USA, with revised intercoolers, turbos, and intake and exhaust manifolds.

The Evo III arrived in 1995. Again a Japan-only model, the Evo III featured a distinctive front bumper, which inspired a million cheap body kits. The rest of the bodywork was new as well, with a larger wing, wider flares, aluminum hood, and other improvements. The engine was heavily revised, although the platform was the same as the previous two Lancer Evolutions. The turbo grew, and the exhaust manifold and compression ratio were both improved. The result was a more than 10-hp increase to 270 hp. The torque remained at 228 ft-lbs.

As hard as it was to imagine that the Lancer Evolution III could be improved upon, Mitsubishi did exactly that with the introduction of the Evo IV in 1996. The car became heavier, but significant engine and drivetrain changes broke the link between the Evo and the Eclipse still being manufactured and sold in the USA.

The engine was still a 4G63t, but it was reversed in the chassis to hang on the right side of the transaxle, rather than the left. This required a new head design to swap the intake and exhaust ports, and numerous external changes to the manifolds, turbo, and plumbing. The turbo grew again, too, to a twin-scroll model, which helped pump output to 276 hp and 260 ft-lbs of torque. The greater power was somewhat offset by the car's heavier curb weight, but one of the benefits of the greater weight was an available Active Yaw Control system that varied torque between front and rear drive wheels.

The Evo III was a revelation to the few people lucky enough to drive one. Recaro seats and a Momo steering wheel were both standard equipment, and the car was capable of sub-5-second 0–60-mph times. Top speed was nearly 150 mph according to contemporary road tests.

Unlike previous Evos, 10,000 Evo IVs left the factory. Some of these were RS models with lighter body panels and glass, and special stripper interiors designed to be an off-the-shelf rally car. Thousands of them were entered into all levels of amateur and professional rally and road-racing competition.

The Evo V arrived on the scene in 1998. It was based on the Evo IV, but redesigned to accommodate new World Rally Car (WRC) rules that reduced homologation requirements and gave them a freer hand in improving the car than in previous years. The body kit was redesigned slightly to allow for wider, less-offset 17-inch wheels surrounding new Brembo-sourced fixed-caliper brakes. A new adjustable spoiler increased high-speed down force. Under the hood, new lighter pistons raised the rev limit, and larger injectors (at 580 cc/minute) gave the engine a little more performance headroom for tuning. The cams and turbocharger were also slightly tweaked, increasing torque to 275 ft-lbs. Horsepower likely increased, but Japanese advertising standards prevented Mitsubishi from noting that fact; the published number was again 276 hp (280 PS).

In 1999, Mitsubishi announced the final Group A-eligible Lancer Evolution. The Evo VI was basically

The Evo III boasted the most highly tuned, 270-hp variant of the left-mounted 4G63t engine. Many parts from this engine can be used to upgrade a USA-market DSM engine, particularly the 2g.

Based on the redesigned and improved 1996 Lancer platform, the Evo IV was the first Evo officially exported to Europe. The most noticeable change to the exterior of the car was the updated Lancer looks, different front bumper, wing, and taillights. (Image courtesy MMNA)

the same as the Evo V with some small tweaks. The front bumper and body kit was redesigned, as were the spoiler and wheels. The lower suspension arms were made from aluminum alloy, and there were a few other suspension and brake changes.

Under the hood, the turbocharger changed yet again, to a titanium-aluminide turbine wheel on the RS model, the first production titanium turbo in a production car. Yet lighter pistons, and a larger intercooler and oil cooler helped boost the spread of power, but the engine's output remained the same as the Evo V. A special RALLIART-tuned edition was available in the UK with 330 hp on tap.

Starting with the Lancer Evolution VII in 2001, Mitsubishi abandoned the practice of building a Group A rally car. The Evo VII was free to become a more powerful (as well as heavier) car in succeeding generations. The car, while similar to previous Lancer Evos, was based on the larger Lancer Cedia platform. Mitsubishi naming conventions between different markets can be confusing; suffice it to say that the

Lancer Cedia was a larger, more-refined car than the Lancer upon which the previous Evos were based.

The Evo VII saw the first use of an active center differential, along with other revisions to the drivetrain. An automatic-transmission Evo VII was even produced for some markets. The engine was only slightly revised from its Evo VI specifications; with torque growing to 284 ft-lbs. The advertised horsepower of 276 (or 280 PS – 1 HP = 1.014 PS) remained the same, as did overall performance.

The Evo in America

In 2003, Mitsubishi launched the fastest, most highly developed Evo yet. The Lancer Evolution VIII sported such high-tech features as a Super Active Yaw Control (AYC) and a 6-speed manual transmission. The AYC and other electronic wizardry managed the flow of power to all four wheels, giving it even more impressive performance than the Lancer Evo VII. The engine was more or less unchanged from the

The Evo VIII (right) and Evo IX (left) finally arrived in the USA market starting in 2003. Unfortunately for USA enthusiasts, we did not get all the high-tech gadgets found on Japanese and European-market cars, including Active Yaw Control, Active Center Differential, or titanium/ aluminide turbocharger wheels.

The ultimate 4G63t is the one installed in this car: the WRC05 Lancer. For its final season of WRC Rally Racing, Mitsubishi and its tuning arm RALLIART pulled out all the stops. Almost no parts are shared with the production engine, including the dry sump oiling system, block main girdle, and internal parts—this was a pure race car.

Evo VII, although special pumped-up versions became available in some markets, particularly the UK. In fact, the UK market saw one of the most powerful production vehicles anywhere: the limited-production FQ400 with 405 hp.

The biggest news for enthusiasts on this side of the Atlantic was the launch of the Evo VIII in the USA. The Subaru Impreza and Audi S4 had awakened enthusiast appetites for all-wheel-drive turbo sedans, and the Evo VIII arrived to compete head-to-head with these cars. While the first USA-market Evos were impressive, they lacked some of the high-tech goodies found in the home market. The AYC was gone, as was the limited-slip front differential and 6-speed transmission, at least for the first year. The engine was the latest-spec 4G63t, slightly detuned for USA smog regulations, but still producing 271 hp and 273 ft-lbs of torque.

In 2004, USA buyers got the RS model with a front limited-slip differential, and in 2005, Mitsubishi

Mitsubishi Owners' Day is a once-yearly event held at Mitsubishi Motors North American headquarters. Mitsubishi drivers and fans from all over the country converge on Cypress, California, for the event. Clean, older, 4G63t-powered cars make up the bulk of the attendees.

Look carefully at the exterior of the car. If it looks clean and well kept, it's more likely to be in good mechanical condition. Also, bodywork and upholstery are among the most expensive parts of any car project, so find the straightest body you can.

launched the MR edition with the 6-speed transmission and active center differential. The 2005 model also gained a few horsepower, now rated the same as the Japanese model at 276 hp and 286 ft-lbs of torque.

The last 4G63t-powered Lancer Evolution was introduced in 2005 as the Evolution IX. Most of the car was carryover from the previous generation, though a few new models appeared. The 4G63t in its final iteration gained Mitsubishi Innovative Valve-timing Electronic Control (MIVEC), a system that varies valve timing on the engine's intake side. This gave the engine a greater spread of useable power on the road. Rated horsepower grew to 286, and torque to 289 ft-lbs. In Japan, the Evolution IX was available as a station wagon, but not in other markets.

Finding and Buying a Turbocharged Mitsubishi

Most of you reading this book probably already own a turbo Mitsubishi, probably a 1g or 2g Eclipse/Talon/Laser, since they're the most

The same rules go for the interior. Look for a clean, tidy interior without excessive wear. Look at the gas and clutch pedals to see if the wear looks like it matches the mileage. As these cars age, interior parts have become quite rare and hard to find.

common in the USA. However, if you don't already have one, or you're looking to grow your collection, a little research will help you find a car that won't turn into a time bomb the first time you try to up the boost.

The first time you look at any potential project, try to get a good idea of its general condition. All of the 4G63t-powered cars tended to be

bought by people who planned to use all their performance. That means that many of them were run hard, and may even have been abused. The early cars' low current value and attractiveness to ham-fisted hot-rodders also results in a lot of trashed cars on the market.

All of these cars have a reputation for poor reliability among some people, but that is likely caused by a lack of maintenance rather than poor build quality. The 4G63t-powered cars suffer badly from lack of maintenance and do not tolerate long-term abuse and neglect. As with any turbocharged DOHC engine, oil changes and timing belt replacements are crucial to the engine's health. In general, stay away from an abused car unless it is very cheap. It is well worth it to pay top dollar for a car with a complete set of service records since the cost of neglect can add up quickly.

That said, the 4G63t is a tough little engine. There are a few problems, particularly with the bottom end of some 2G engines, but the engine is as durable and reliable as

The 4G63t in its native habitat. This one is in a Galant VR-4, one of the first 4WD super-sedans available in the USA. It's a great car, and with a few well-chosen modifications, it can hold its own even against the latest Evo IX and competitors. You want to find a car with an engine that looks like this, but they're few and far between.

any Japanese car of the time. As long as a car is properly maintained and not abused, there is no reason a 4G63t drivetrain won't last more than 200,000 miles without major surgery.

The first thing to check on any turbo engine is blue smoke from the tailpipe. High-mileage engines have a reputation for burning oil, which is generally caused by worn valvestem seals. Valve seals will fail around 120K miles, causing smoke, although they can fail as early as 60,000 miles if the oil is not changed often enough. The latest parts supplied by your local Mitsubishi dealer are better than the original seals, and should last for the life of the engine.

Another source of oil in the combustion chambers is the turbo. The turbo installed on most 4G63t engines is a high-quality Mitsubishi unit with a water-cooled center section. Turbos can last 150,000 miles with proper care, which means frequent oil changes, but failures are common if the engine is not maintained. A bad turbo can cause smoke under acceleration, or, in the worst case, a scraping noise when the boost comes on. If the engine is burning a lot of oil because of a worn turbo, budget for a replacement.

Burning oil can also come from worn rings, but this is not usually an issue unless the engine has 200,000 miles or more. In that case, you're guaranteed to have to pull the engine and rebuild the bottom end. Before you buy any turbo car, check the compression—it doesn't take long and it will tell you right away if the engine is in passable shape. Of course there are a lot of other problems that won't show up in a compression test, but it is a good quick check for a bum motor.

Watch for white smoke, as well, and check for any signs of coolant in the oil. Either symptom points to a bad head gasket, cracked block, warped head, or worse trouble. An engine that looses coolant is not likely to last long, and has probably been abused. Check also for a recently installed thermostat or radiator. These parts are often changed by people trying to solve an overheating issue that may have already damaged the engine.

Listen to the engine start from cold, if possible, to hear any noises clearly. Any knocking, especially if it is at the same speed as the engine, is cause for concern. Top end noises are less critical, but they should still be

The good thing about valvestem seals is that it is a relatively easy fix, though it does require a tear-down or possible removal of the head, which means replacing many other wear parts is a good idea.

Look under the coolant filler cap for signs of oil in the coolant. A rusty color is perfectly normal. Change the coolant on any car with an unknown history and check carefully for other cooling system damage since this is one of the weak spots of the early cars.

investigated very closely to make sure that something expensive isn't damaged, like wiped cams or damaged followers. Check the condition of the oil. A nice honey or brown color is optimal; anything darker is an indication of oil change intervals that are too long. Smell it for fuel contamination, too. If the injector seals are leaking or there is some other fueling problem, gasoline will get into the oil, thinning it out and causing bearing problems in the lower end

Galant VR-4 and 1g/2g DSM

The earliest 4G63t-powered cars are fast approaching the 20-year-old point, which means that many of the problems common to any old car are an issue. Many of these cars have reached 120,000 to 150,000 miles, so only the best-maintained examples are even worth considering.

Many 4G63t engines have noisy valvetrains, which is not a big deal and also easy to fix. In the best-case scenario, the noisy lifters are caused by infrequent oil changes and cheap oil filters; an oil change and flush and a new filter can completely cure

These little round components are the ECU capacitors that often fail on DSMs. They won't give much warning before they fail. If you hear clicking under the dash and the engine runs erratically, the ECU is the most likely culprit.

Older DSMs have wiring harnesses that are nearing or beyond 20 years old at this point. Many harnesses are no longer available, so check yours carefully for signs of damage. The connectors, like this one at the throttle position sensor (TPS), are the first to go. Pins and some shells are still available to repair connectors.

some engines. In the worst case, the lifters have completely collapsed and must be replaced.

Like many turbo engines, cracked exhaust manifolds are a frequent problem with the early 4G63t. Luckily, Mitsubishi noticed the problem too, and later manifolds are not as crack-prone.

The fuel-injection system is durable, but leaky injectors are an issue as with most cars of this age. Injector troubles are more common in states like California that use a cocktail of fuel additives. The most common failure of the EFI system on the early cars is leaky capacitors in the ECU. Over time, these small electronic parts have a tendency to leak acid on the PC board and damage it, which kills the ECU and stops the engine dead. Several vendors offer a capacitor-replacement service, although the repair is not hard to do if you are handy with a soldering iron.

The connectors and harnesses on every part of the EFI system are also likely to be brittle by now, so be careful disassembling them, and check any prospective purchase for broken

connectors. They can all be replaced, but harnesses have become hard to find, and replacement connectors are not always available. It's also no fun to repair ratty wiring harnesses, so start with the best one you can find. Also, save connectors and other parts you find in the junkyard to repair broken connectors as they occur.

There is a problem that is seen sometimes on the 2g Eclipse/Talon, the well-known "crank walk" problem. The causes and severity of this problem are a little vague, and even the most experienced Mitsubishi tuners disagree on the specifics. What they do agree on is that some 2g Eclipse/Talon cars have had problems with the crankshaft thrust bearing.

Check any 2g DSM (they all have 7-bolt engines) and 1g DSMs built with a 7-bolt engine. How do you know if your engine is a 6- or 7-bolt? The easiest way to check is reach down below the front pulley (with the engine off, of course!) and feel the rail between the engine and oil pan. 7-bolt engines dip down towards the oil pan. Six-bolt engines, on the other hand, have a flat oil pan rail. Look at the pictures on page 20 for clarification. A good test is to press the clutch pedal while the engine's running and watching for any movement at the

This is the upper half of the center main bearing. It contains the thrust surface that wears out and causes "crank walk" in some 7-bolt engines. This one is new, but will eventually become worn from the force of the clutch and accessories.

crank pulley, or any strange noises that vary when the clutch is depressed. Check to make sure the clutch returns properly to neutral.

At the very least, crank walk means you'll need new bearings and likely a new crankshaft, but it can also mean that the block, oil pump, and front cover are trashed. Either way, this is an expensive repair, and should automatically disqualify a car from consideration unless it is very cheap.

USA-Market Lancer Evolution

At the time of this book's publication, USA-market Lancer Evos are still fairly new, although some are getting on in miles. Check the same things mentioned above—any noise, smoke, or other trouble is a sure sign of an abused engine this new. Some Evo engines are known for having noisy lifters, but crank walk is not usually an issue, and the cars are not old enough to have the same electrical problems that afflict the earlier cars.

Do make sure that the car has full service records, and look carefully for any signs of abuse. These are highly tuned cars and many drivers don't have quite the skill they think they do. Burnt clutches and hard shifting are sure signs of rough treatment. The only real weak spots of the latest Evo are the clutch and transfer case—the transmission and engine are tough.

Engine Swaps

One last thing to watch out for is engine swaps. As these cars get older, more and more engines are swapped between chassis, confusing the issue of parts interchange and specifications.

There is nothing wrong with a car that has received an engine, head, or long block from an earlier or later year, but you should know about it before trying to order parts or modify anything. At a minimum, make

Few cars inspire enthusiasm like the USA-market Lancer Evolution VIII and IX. Buy your car from a fellow enthusiast and your chances of buying one that hasn't been abused are much higher. Check out the "for sale" section of websites and forums, but don't get so excited that you overlook major defects.

sure you know if the engine is a 6 or 7 bolt, 1g or 2g engine, and if it's got the right head and intake manifold for the block. It also helps to know what engines are better from a performance or modification standpoint so that you can look for the right parts to build your own motor.

While the 4G63t and 4G64 engines are all very similar, there are enough differences in engine controls, accessories, and mounts to make direct swaps a hassle. DSM motors can be swapped from 1g to 2g, and from Galant VR-4 to 1g or 2g pretty easily.

Impossible swaps include those from USA-market Evos to any DSM. First of all, the USA-market Evo motors (including every Lancer Evolution from IV on) are rotated 180 degrees compared to the USA-market DSM motors. The timing belt and accessories are on the left (USA passenger) side of the engine compartment. This means the mounts, covers, and accessories are totally different.

Other uncommon 4G63t swaps you might come across are those into various Hyundai models, the Mirage

Turbo, and Starion. These swaps are mostly outside the scope of this book, but they are not impossible. Any car originally equipped with a Sirius family engine can accept a 4G63t with enough hammering and engine mount fabrication. Make sure you use the right block for your transmission (narrow FWD or wide RWD bolt pattern) and you should be golden.

1g into a 2g

This is the most common 4G63t motor swap because of 7-bolt crank walk problems, and because the 6-bolt engine has stouter rods and crank. It became popular when these cars were newer and the 2g engine was significantly more expensive, although with today's aftermarket parts and careful assembly, it's no big deal to get the same power and reliability out of a 7-bolt engine as a 6-bolt. If you want to perform one of these swaps, rest assured that the path has been well paved since the mid 1990s.

The swap is easy from a bolt-in standpoint; the 1g engine mounts on the front and rear of the block are not used, since the 2g mounts are located on the transmission. You must use a 2g driver's-side engine mount bracket on the 1g engine's front cover. This requires slight modification to the engine mount. Just trim the corner that fouls on the water pump with a grinder. Trim only as much as necessary for clearance, going a little at a time.

Neither the 1g nor the 2g timing cover will fit the hybrid engine properly, though the 1g is the closest to fitting. Modify the mounting holes as necessary to fit one on—don't even think of running without it. Robert Garcia likes to safety wire the rear of the cover so that it does not flap loose. Don't forget that the new motor will also need a 1g flywheel, but the clutches are interchangeable. The 2g starter and starter bracket will work with a little modification, and you'll have to use the 2g accessory brackets, including the power-steering bracket.

This swap also requires changes to the engine control sensors. The 1g has only a camshaft angle sensor (CAS) while the 2g has a CAS and a crankshaft position sensor (CPS). The 2g CPS is mounted in the oil pump housing on the front of the motor, which is designed very differently from the 1g oil pump housing. The 2g CAS is also different (or at least the 1995–1996 one is) and it mounts under the cam pulley on the intake side.

Use a 1g CAS; it already outputs the proper signals for the 2g ECU to run the engine. You will have to make or buy a jumper harness to go between the 2g engine compartment wiring and the 1g sensor, or you can just use the 1g sensor and hard-wire the connection. Just make sure to connect the right wires and you should have no problems. Use all the other 2g sensors, including coolant temp, TPS, knock sensor, etc.

Evo Swaps

There are a few Evo VIIIs running around with Evo IX engine swaps. While the MIVEC head is better than the earlier one for power production, it probably isn't worth the expense and trouble of swapping it into the earlier Evo.

If you must do this swap, it's a very straightforward one from a mounting standpoint since the two cars are nearly identical under the hood. The biggest differences are electronic; make sure that the wiring harness matches the engine you're installing and you shouldn't have a problem. To get the most benefit from the IX engine, be sure to get the intake plumbing and turbo, since there are minor differences.

If you're determined, this engine can be swapped in place of the 4G64 installed in the 2000-and-later Galant and Eclipse, as well as the 2.4-liter 4G69 installed in the most recent Eclipse, Galant, and 2003–2006 Lancer and Outlander. These engines have the same orientation and bellhousing bolt pattern as the Evo, although there are many differences between the cars that make this swap a difficult one. The mounts might be easy, but you will have to design your own intake plumbing using parts of the stock Evo plumbing, and you'll have to rewire the donor wiring harness to mate with the car's harness. You might also find that the stock transmission isn't up to handling the power of the turbo engine, but that shouldn't stop you if you've made it this far!

4G64 Swaps

The 4G64/G4CS engines are based on the 4G63 engine block and so they are a straight bolt-in for most 4G63t cars. The reason for this swap is the block and crank alone—with a 4G63t head and turbo you've got an engine with 20 percent more displacement and nearly 20 percent more power potential than a 4G63. The larger displacement will help spool a big turbo, and give you extra useable street power at lower RPM.

The key to this swap is the DOHC head and turbo. The DOHC head shares the same bolt pattern as the SOHC head, although there may be some oil drain passages on the block deck surface that have to be plugged to use it, depending on the year. Of course pistons with the proper valve reliefs have to be used, but the stock ones should be replaced for turbo use anyway. Depending on the year, it may not have a properly tapped boss for a knock sensor, as well as turbo oil and coolant feed fittings. All of these problems are easily overcome.

The easiest hybrid DOHC 4G64 swap is into a turbo 4G63t-powered car. The 4G63t electronics, intake manifold, and fuel system can be used as-is, and the engine will look identical to the stocker from the outside. If you want to swap your turbo hybrid into a 4G64-powered car, be prepared to swap engine management and fuel systems to fire the DOHC's coil packs and keep up with the fuel demands of 250+ hp. It's not a hard swap, but it's not as easy as dropping a 2.4-liter hybrid into a 4G63t car.

Buying a Core Engine

There's one more topic that deserves a good look at this point, and that's buying a used core engine. In general, treat any non-running engine as a core block that will have to be rebuilt. No matter what the seller says, if you can't hear it run you can't determine if it is worth using as-is. Swapping an engine is a lot of work; nobody should have to go through it twice because the first engine was no good.

DSM engines come in two varieties, 6-bolt and 7-bolt, so named for the number of bolts holding the flywheel to the end of the crankshaft, though there are dozens of other differences. The 6-bolt engines are earlier and in some circles considered better. The 7-bolt engines are those known for possible crank walk issues.

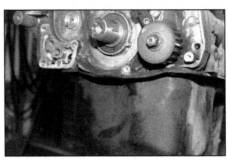

This little dip in the oil pan rail is a quick way to check if you have a 6-bolt or 7-bolt engine. The oil pan is flat in this area on a 6-bolt engine, which means this is a 7-bolt core. There are other ways of telling the two engines apart, but this is one of the quickest if the engine is still in the car. You don't even have to lift the car to check.

There are a couple of exceptions to this rule. The first involves any engine that you thoroughly inspect. First, pull the valve cover and turbo before you buy it. If the aluminum under the valve cover looks clean and silver, that suggests that the engine has had regular oil changes. The turbo exhaust turbine should be evenly dull brown or gray. A black, gooey coating on the turbine blades means the engine has been burning a lot of oil. Check for oil on the cold side of the turbo, too; oil inside the compressor housing is from a bad turbo bearing.

If you can perform, or have a shop perform, a leak down test, you can tell if the ring and valve sealing is good or not. None of these tests will tell you about the condition of the bottom end, oil pump, and bearings; so to be absolutely sure you're getting a good used engine, pull the oil pan and check the thrust bearing and rod bearings.

There are a lot of parts suppliers advertising "Japanese Import" engines. The companies that import these engines claim that they have only 30,000 to 60,000 miles on them, which may be true, but it does not mean that they're automatically in good condition. It may not even be true. What can be said about them is that, for the most part, Japanese driving patterns are very slow stop and go traffic with minimal long trips.

The engines that come from Japan can be in good condition but they will suffer from the same treatment as USA-market engines from the junkyard. Only buy Japanese engines with a guarantee, and only if you can perform some of the tests listed above. Many importers have also started performing leak down tests on their imported engines; go with the most thorough supplier, if you can.

Make sure that the engine you're getting is compatible with your car or be prepared for some small hassles. Note whether the engine you're getting is FWD or AWD, and if it includes the flywheel or flex plate to match your car. Most of these engines are the same as their USA-market counterparts but be ready to deal with differently sized injectors,

smaller or larger turbos than you expect, and accessory differences.

Don't fall for the trap that the so-called JDM engines are much, if any, better than the USA-market engines. For the most part, the 6-bolt 1g engines are all the same, as are the 7-bolt 2g and Evo engines. Some of them had larger injectors, the so-called yellow top 510-cc/min injectors, and larger "small-16G" turbos, but these engines are rare.

Some of the engines from Japan are marked "Cyclone" on the valve cover or intake manifold. These engines simply have a second set of intake runners with an actuator and set of butterflies. They're a way of boosting midrange torque by tuning the intake system similar to Toyota and Nissan systems. Unless you have the matching ECU and wiring, there's no way to actuate them on a USA-market vehicle, and no real benefit to making it work.

Maintenance

Once you've got your Mitsubishi turbo, take the time to catch up on neglected maintenance if you want to modify it. Regular maintenance includes oil changes and coolant changes; see the note above about what happens when you neglect either one.

Mitsubishi recommends replacing the timing belt every 60,000 miles, and this is not a service that should be overlooked. A broken timing belt will result in bent valves in the interference engine. The balance shaft belt is just as important; most timing belt failures are caused by a broken balance shaft belt. Because of the way the engine is laid out, there are a lot of small parts that should probably be replaced at the same time as the timing belt, including the crank and cam seals and timing belt tensioner.

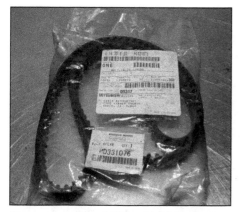

Use only original Mitsubishi timing belts every 60,000 miles. Robert Garcia recommends using only factory timing belts because of their superior heat resistance and quality control. Aftermarket belts may be fine, but the price difference is small enough that using an OEM piece is cheap insurance.

On every second timing belt change (every 120,000 miles) replace the water pump, since it is behind the timing belt and usually wears out after 100,000 miles or so. You will have to pull the timing belt again if it goes out, so replace it as a preventive measure.

Every second belt replacement (at 120,000 miles) should definitely include replacement of the idler bearings, tensioner, water pump, and the small hoses that feed cooling water to the turbo.

It's not really maintenance per se, but you should also check the ignition timing after changing the timing

The timing belt idler bearings should be replaced every 120,000 miles even if they don't show any signs of wear or roughness. Always use OEM Mitsubishi parts for the same reason as the timing belt. If you shop smart, you can find them for only a few dollars more than aftermarket.

The 4G63t engines have a reliable hydraulic tensioner. However, it too should be replaced every 120,000 miles. A failed tensioner will cause the belt to go slack and skip, destroying valves, valve guides, cam followers, and even camshafts.

belt. Also check and reset the idle and other EFI adjustments.

Power Goals and Budgeting

Step one of the process of modifying an engine is deciding on a power goal. Before you turn a single wrench, you should have an idea of what you're expecting from the finished engine. This decision will form a framework for many of the things you do later, and will influence which parts you should be buying as you put the engine together. This is also important if you're just freshening or hot-rodding your existing 4G63t instead of building one from the ground up. A power goal will keep you on track.

So where do you get this information? The best source probably is not from reading car magazines and "bench racing" with your buddies at the track or online. Yeah, you might get an idea of what you can do with your car or engine, but most people have unrealistic ideas of what power their engine makes and what they can afford. Even the factory ratings are a little optimistic, but they are a good starting point. The best way to go about the process is to look at dyno charts and the information in the "Bolt-ons" and "Tuning" chapters of this book, or talk to experienced Mitsubishi turbo engine builders. If you ask nicely, any professional engine building shop, like Road/Race Engineering, will give you a rough idea of what you can expect from your engine and budget.

Chassis Factors

Your power goal should be tempered by factors outside the engine. For example, you should consider how your car's chassis and suspension will be impacted by the increased power. Brakes designed to stop a 195-hp sports coupe will probably be heavily taxed by a 450-hp race-tuned engine. This includes the suspension and drivetrain design—a front-wheel drive turbo Eclipse is going to have a tough time putting even the stock 200 or so horsepower down in first gear.

The 1g and 2g transmissions, axles, and differentials can also give you problems. The early stock rear axle (the so-called 3-bolt axle) is one weak spot, as is the transfer case. 400 hp is fun and all, but only an Evo drivetrain can handle that kind of power on a regular basis without breaking something (usually a differential or axle).

Tire clearance and thus maximum size is another factor influencing your ultimate power goals. If you can only squeeze 6-inch-wide tires under your rear fenders (as on the 1g DSM), don't expect to put down much power without vaporizing them.

In short, don't expect to be able to build a killer motor without upgrading other parts of the car—every part contributes to the ability to handle extra power and put it to the ground efficiently. This goes double for maintenance. If the rest of the car isn't up to snuff, spending money on your killer motor might just be throwing good money after bad; use that money to make sure the rest of the car is in good shape first. The 1g and 2g DSMs are getting so old that it is likely most of the running gear and electrical systems are worn out and will need a lot of attention first.

The "Bigger is Better" Trap

When planning your dream engine, it's hard not to fall into the trap of thinking that only the ultimate in power will do. Sometimes, as hard as it may be, it is better to take a step back and plan something a little more realistic. Sure, it would be fun to have the bragging rights that a 500-hp, 2.4-liter monster with a huge turbo brings, but the maintenance and tuning demands of such a beast would quickly become tiring. Also, getting big power out of a small motor like the 4G63t means lots of boost, lots of hours of tuning, lots of

You can rebuild an Evo long block like this one for about $750 in parts and $500 in machine work if you use mostly OEM parts and keep the custom machine work to a minimum. If you want to use forged pistons and rods, along with stronger main and head bolts, be prepared to double both figures.

revs, and a powerband that's almost useless on the street. Keep your goals manageable and useable (see previous note about drivetrain) and you will be happier with the outcome.

Budgeting

Probably the biggest consideration for most of us is the total cost of our project. Without trying, a maximum horsepower late Lancer Evolution engine could easily cost $10,000 or more (much, much more). Unless you're in an unlimited-modification race class, and sometimes even there, you need to keep in mind what you can afford. It goes back to the old and cheesy cliché, "Speed costs—how fast can you afford to go?"

Balance is a point that we make in a couple of places in this book. You can make a lot of power with a DSM without spending a lot of money, but you can't generally have reliable power without spending a lot of money.

Answers to the question of "How much will it cost?" can vary significantly. If you do the labor yourself, you could build a 1g or 2g bottom end for $1,500 including machine work, even less if you can recycle some used parts, and about double that if you use forged pistons with stock rods. Add in a set of rods and hardcore hardware and you can burn up most of $4,000. The head is where costs add up the most quickly. Figure $500 for a stock rebuild with used cams and valves but new guides and machine work. The sky's the limit when you start including things like aftermarket cams, adjustable cam timing pulleys, solid lifters, big valves, and the rest. Porting is also pretty expensive; home porting should be limited to port matching, so budget $1,000 or more for a professional job.

Also, with turbo engines, the turbo, intercooler, and plumbing can be a significant investment—from $1,000 for mostly stock rebuilt parts to more than $3,000 for all-new everything.

Consider your budget before you start and make sure that you can afford what you set out to do. If you can't build your dream Evo motor all at once, think about building an engine in stages—the bottom end first, and then a head swap in the car, and finally a new turbo and intercooler. Another approach is buying a core engine and building it up slowly, over time. Either strategy will let you spend what you can afford as you have the money. This isn't a bad way of approaching a project on your daily driver regardless of cost; it minimizes the time your car will be cluttering the garage or driveway.

Intended Use

One more thing that you should consider during the planning stages is how you plan to use the engine and car package. Is this going to be your daily driver? A weekend fun machine? A rally, drag, or autocross car? Think long and hard about this one, because it will influence many of the choices you make.

The easiest car to build is a race car, because all you have to do is take a look at the rules, figure out the best combination of allowed parts and modifications, and build your engine. Since the operating conditions are known, the engine can be very carefully tuned to perform best in those conditions. In addition, most race cars are trailered to events, so you don't have to plan for street driving or street legality. The 4G63

Building a pure race car is easy. Expensive, yes, but not difficult: Simply replace every part that might fail with something better that fits within the rules. Then go racing until something breaks, tow home, and repeat. 4G63t-powered cars make excellent amateur race machines because they are so tough.

engine has been used in many different racing classes and continues to be competitive in some of them, so ask around if you plan to build a race car and figure out what is being used in your class by the top racers.

The daily driver street car is the hardest one to build, because its normal operation (driving you to work and taking road trips) is opposite the conditions listed above for a race car. It has to be reliable, not too

A front-wheel-drive Talon is not most peoples' idea of a good race car, but at least one has been successful. The 4G63t is capable of producing enough reliable power to win consistently, and almost any chassis can be developed with enough money thrown at it.

This Evo is a gutted track car, with many compromises made to run flat-out without concern for comfort. The full-cage and Spartan interior wouldn't be comfortable for most commuters, though some people manage to ignore the discomfort of driving a racer on the street.

If you're tempted to cheat local smog laws either through avoiding the test or bribing a technician, think twice. Modifications to engine smog equipment constitute a felony in the United States, and you could be liable to return the car to stock for any future buyers.

The air in most large USA cities has improved significantly over the last 30 or so years, and you can thank those sometimes-onerous smog laws for this. Removing or modifying smog controls might make your car go faster, but they will also have a detrimental effect on air quality.

Think about it before you run a car without a catalytic converter, for example, or before you modify the evaporative canister or EGR system. In general, a newly rebuilt engine will run much cleaner than the old one that it replaces, so choose your parts to maximize performance without affecting exhaust emissions.

uncomfortable, and able to sit in traffic without overheating or stalling. That's a lot to ask of a highly tuned engine, so it will dictate the parts you choose and the tuning that you will require.

That's not to say that you can't build a really monster daily driver, you just have to be willing to accept compromises such as a smaller cam, or weak off-boost performance caused by the low compression ratio and big turbo needed for weekend fun runs.

Other Factors

Before starting on your engine upgrade path, don't forget to think about your local laws, and the effect any changes will have on your car's resale value or originality. Local emissions regulations and sometimes safety regulations can be an issue for those of us in some states.

This engine wouldn't pass smog in any USA state. There are simply too many modifications for it to be legal. As tempting as it might be to build a race motor for the street, the costs of getting caught in some states are very high.

Other concerns for the budding hot-rodder are the value of the car, and the level of your own skills. If your car is rare (like the USA-market Galant VR-4), make sure you're ready for the consequences of changing it. It's your car, though, so don't worry about offending the purists. Rarely do you get even a fraction of the cost of speed parts back when you sell a car. In fact, many times the modifications reduce the pool of potential buyers and make it much harder to sell.

Finally, try to keep your own resources and skills in mind. If you're not an experienced engine builder, you should have no problem building or modifying your own engine, but you will need to have an open mind and be ready to research lots of details. You'll also need a place to work. You can build an engine in the living room and swap it on the street but it certainly isn't the best way to go about it! Rent a garage if you don't have one and your life will be much better.

Paper Engine Building

Once you've worked out a realistic power goal and a comfortable budget, take the time to build your engine on paper or on a spreadsheet in your computer. Read this book from end to end and take notes on parts that interest you. Look them up in manufacturers' catalogs and put the prices in your dream-build spreadsheet. At first, your spec sheet will be more of a wish list than a plan, but as you do your research you will find that some things won't work the way you expected, and others will work better.

The first draft of your spec sheet should have, at a minimum, the following information:
- Engine original application (Galant VR-4, 1g, 2g, Evo VIII, etc.)

Even a few years after it was first sold, an Evo is a fairly expensive car. Modifying one to build a race car is almost always a losing proposition—it's very difficult to get your money back out of a race car unless you do all the work yourself. Even then you'll not likely make more than minimum wage on the deal!

- Stock power rating (190–282 horsepower)
- Expected power
- Current or expected use of the car
- Your budget

As you near the start of your project, add a list of needed parts, and a list of machine work that needs to be done. To the information above you should add:

- Piston type and brand
- Final bore and stroke
- Rod type/modifications
- Crankshaft type/modifications
- Head work (porting, milling, rebuilding)
- Block work (boring, decking)
- Intake parts needed
- Exhaust parts needed
- Valvetrain parts needed

- Camshaft to be used
- Hardware to be used (head and rod bolts, etc.)
- Turbo and exhaust manifold to be used
- Turbo plumbing, intercooler, and BPV
- Engine management

At this point you're ready to start doing research and collecting parts you need. As you get a quote for a part or machine operation, enter it next to the item. That way you can compare a running total to your budget. When you actually pay for each item on the list, put the actual total next to the budgeted total to keep track of your total costs. It sounds like a lot of work, but it will save you many hassles down the road.

BASIC BOLT-ON MODIFICATIONS

Most of this book deals with the guts of your engine, starting at the turbocharger and intake and going inside from there, since that is where the real power of a 4G63t comes from. But that's not the whole story, and for some people it may not even be part of the story. The most common enthusiast modifications are external to the engine, and involve nothing more exotic than bolting on a few parts.

Some versions of the 4G63t respond very well to these simple external modifications, also known as bolt-ons. They are surprisingly effective at squeezing more power out of an otherwise stock 1g or 2g Eclipse engine, for example. The definition of bolt-on varies from person to person, so for the purposes of this chapter we're going to limit our discussion to only those items that can be installed without removing the engine, transmission, cylinder head, oil pan, manifolds, or valve cover.

Common items like adjustable cam timing pulleys, aggressive camshafts, ported intake manifolds, engine management upgrades, and some other parts are also found under the hoods of some mildly modified street cars, but they require

A few thoughtfully chosen modifications to the outside of your engine can make your car faster and easier to tune later without taking away any reliability. Choose good-quality parts from reputable manufacturers and install them with care and your engine will look like this, and perform better.

significantly more experience and work to install and tune. They will be covered in later chapters.

Expectations

One of the best things about the 4G63t is the strong bottom end and other internal engine parts. While some manufacturers build engines with only a thin margin of strength, the 4G63t can take a lot of abuse without suffering much. The stout DSM short block can easily handle 17 psi of boost. The Evo block is stronger, but fuel octane will usually limit boost to around 20 psi on the street. For short bursts, as with a street car, the DSM motor will handle 325 hp on the stock bottom end in good condition and the Evo motor about 425 before reliability becomes an issue. Of course this assumes a good head gasket seal, proper maintenance, precise tuning, and good treatment; no engine will survive long if it is abused and neglected.

The modifications listed here enough to build an early 4G63t up to around 250 hp, or an Evo VIII or IX engine to around 300 hp with almost no drawbacks. That's nothing to laugh at in either case, representing between 10 and 25 percent more power than stock at about the same level of reliability.

Staged Tuning

As mentioned in the intro chapter, engine tuning should be thought of in terms of an ultimate goal. Rather than a series of unrelated steps, each point along the journey from stock to modified should be thought of as a stair step; the parts you need to reach 300 hp often include parts needed to reach 250 hp.

To help out with this process, companies started marketing tuning in stages, where the parts included in one stage also carry over to following stages. It's a good way to think about any modification, and wherever possible we've tried to stick with that terminology.

To be successful with a staged modifications strategy, take good notes after every change. Drive the car to record both measurable performance improvements and changes to the way the engine responds. This will all be covered more in Chapter 6, but get in the habit of taking notes while you're still thinking about changing things.

Make sure that the timing belt cover is intact and securely fastened. The timing belt is the single most important external part of the engine and a small rock or debris could cause it to skip a tooth or break with disastrous results, like bent valves.

Now that you've got this book in your hands and read all the background, you're probably looking for the part where you get to spend some money and make some power. Well, hold on for just a second. Before you can ask an engine to give out more power, you have to make sure that it is up to the task and make sure it's running right and not worn out. This should be so obvious that it doesn't deserve a mention, but...hey, everyone gets sloppy sometimes, right?

The process of giving an engine a tune-up and setting a baseline is also sometimes called "stage 0" because it has to be done before even the first stage of aftermarket parts can be added. A stage 0 engine has a new or fairly new timing belt, tensioner pulleys, and so-called expendable parts like a water pump and clean cooling system. A good candidate for bolt-on modifications will be clean and won't have any leaks (oil, water, or otherwise). These are all small things, but added together they will make working on the engine easier, cleaner, and safer.

A ready-to-tune engine should have no major oil-burning tendencies, though a little consumption is not something to worry about. Oil reduces the octane of the fuel in the combustion chamber and will limit the amount of boost you can run, but overall it's an acceptable tradeoff when you consider the cost of rebuilding an engine to cure a small oil-burning problem. Even if you don't plan to rebuild your entire bottom end, you should take the time to make sure it is still in good shape before attempting to tune your engine, increase boost, or perform any serious modifications.

Running right, from a performance standpoint, also includes fresh copper spark plugs in the factory heat range (BPR8ES for DSMs) and an ignition system that isn't lacking in wattage. All that high-pressure air in the combustion chamber requires lots of power to ignite consistently. Just make sure that the plug wires and spark plugs are new and not fouled from previous problems, and make sure that the coil pack connectors are clean, tight, and unbroken.

If your plugs are new or in almost-new shape, check your plug wires for conductivity and condition. If you don't know when they were replaced, or it has been more than a few years, replace them with good-quality replacement wires like these NGKs.

On the fuel side, maintenance includes making sure the fuel pump isn't starving for fuel and that the injectors aren't clogged. If in doubt, replace the former with a quality aftermarket replacement like the Walbro 190 lph pump, and send the latter to someone like Road/Race Engineering for a dip in the cleaning tank and a trip to the flow-equalizing bench. These parts may be upgraded later, but tuning the first few modifications will be easier with known good parts installed.

Check the ECU for bad capacitors and make sure all the under-hood connectors are tight and clean. Check and fix any codes that the ECU gives and make sure that they don't recur. Set the ignition timing to the factory settings using your factory shop manual (you do have one, don't you?). This can only be done if you have a stock 1g DSM or a 2g with an adjustable cam angle sensor (CAS). Make sure to short the timing connector to ground so that the ECU holds timing steady while you adjust the CAS.

The 4G63t ignition systems are pretty good. Even engines making 400 hp don't generally need much in the way of ignition help. The coil packs are reliable and produce a good spark. The only weak links in the system are the connectors, since they are getting so old, and the high-voltage ignition wires. It pays to get the highest quality ignition wires you can find. Some of the aftermarket "super" wires are not as good as stock wires, so be sure to shop around for wires that are high quality, like stock replacement NGK wires. Don't expect much in the way of power increases, though you may see a one or two horsepower increase on the right engine on the right day. Wires and plugs will be covered in more detail in the ignition chapter.

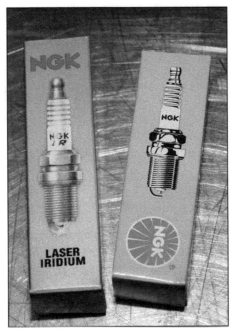

Get a set of new stock plugs before tuning your engine. Use the OEM iridium NGKs in an Evo 9 head, and copper BPR8ES for everything else. Iridium plugs last a little longer, but copper plugs are a good solid choice for a performance turbo engine.

Check and fix any boost leaks—no sense tuning the engine to produce more boost if you're just going to loose it out of leaky plumbing and intercoolers. It takes a lot of horsepower to pressurize air, and if it is blowing out of a leaky pipe or seal then you're just stressing the engine unnecessarily. As these cars age, the gaskets and seals in the intake harden and begin to leak air first; the result is rough running and boost leaks.

One of the last things to get in stage 0 is a tank full of good fuel. The 4G63t in all its variants is a high-performance engine, and it does not like fuel with low octane. Make sure you have the right fuel in the tank at all times or you're giving away very cheap horsepower. The right fuel has enough octane to prevent knock. This is especially important if you're in a 91-octane state like California.

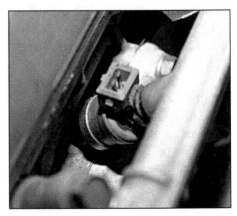

The injector seals in the manifold are common sources of boost leaks on DSMs. Also check the gasket between the throttle body and manifold, and the gasket between the intake manifold and head. Any leaks mean the parts have to be pulled and gaskets replaced.

These engines use a knock sensor that retards timing when any knock is detected, and that retarded timing costs power, so make sure you have the best gas you can get. Anything less than 91 octane is just testing how well the knock control works. If the engine knocks (or retards timing) on 91 octane, you're going to have to pick up a few extra octane by adding octane booster or topping off the tank with "race gas" of 100 octane or better. You'd be surprised to know that a stock DSM can gain as much as 5–10 hp simply by running on 104-octane race gas.

Pre-Tuning Engine Checklist

1. Check oil pressure and listen for bearing noise. Inspect 7-bolt DSM engines for crank walk, pull the oil pan, and inspect bearings if necessary. Rebuild if any unusual wear is found.

2. Check compression and oil burning; replace valvestem seals or rebuild engine if necessary.

3. Check ECU and all connectors for damage; replace capacitors and re-solder connectors if any are loose or broken.

4. Catch up on any maintenance—replace the timing belt, seals, tensioners, and water hoses if longer than 60,000 miles since last replacement. Replace water pump if 120,000 miles or more.

5. Check for and repair any oil or coolant leaks.

6. Check cooling system for clogged radiator, bad fan motors, or leaking water pump.

7. Check turbo; replace or rebuild if bad.

8. Change oil, filter, and coolant.

9. Check for cracked exhaust manifold and replace if necessary, especially on 1g DSMs and Galant VR-4s.

10. Check for boost leaks and replace damaged intake hoses, gaskets, and injector seals—check the PCV valve system for leaks, too, and replace it if you don't know how old it is.

11. Pull injectors and send out for cleaning and balancing.

12. Check fuel lines for leaks; replace fuel lines and filter.

13. Replace fuel pump if noisy or more than 100,000 miles since last replacement.

14. Install new copper spark plugs and wires if more than 30,000 miles since last replacement.

The stock DSM boost gauge is driven by the ECU based on airflow and boost controller position, and does not reflect actual boost, especially on a modified engine. Use an aftermarket gauge if you want to know what's going on.

Gauges and Instruments

Along with taking good notes, you should also keep track of the health of your engine so that you don't break anything. The stock gauges aren't going to work very well for this. Depending on your vehicle, you've got only an indirect boost gauge and either no oil pressure gauge or an inaccurate stock gauge. No Mitsubishi turbo engine comes with a way of keeping track of the air/fuel mixture in the engine, either. All three of these data points (boost pressure, oil pressure, and air/fuel ratio) are important for making horsepower with a healthy engine.

Boost Gauge

You should add a boost gauge as soon as you start to modify anything. Since boost is closely tied with both horsepower as well as broken parts, you will need to keep a close watch on the real boost as you change things.

Mechanical gauges are the cheapest, and can be very accurate. However, they also can suffer from friction in the mechanism and may lag behind actual boost in the manifold. The length of the tube that directs boost pressure to the gauge adds to its delay. Electrical boost gauges, on the other hand, are more expensive but do not have the same trouble with delay. The sender can be mounted directly on the intake manifold with an adapter, or near a boost source. That makes them more responsive to fast-changing boost conditions.

A good quality boost gauge is a necessity, too, since knowing the manifold pressure is only helpful if it is accurate. The difference between 14 psi and 16 psi, based on your engine, fuel, and timing, could be the difference between driving home and calling a tow truck. Don't be tempted by cheap parts-store gauges that look nice but aren't close to accurate.

A boost gauge can be complemented by an accurate oil pressure gauge to keep an eye on the bottom end—good oil pressure is critical for

The latest trend in gauges is digital displays that can be reprogrammed to display whatever information you want. The AEM gauges are an excellent example—the large steady numbers can be easier to read at a glance than vibrating analog needles.

reliability. Oil pressure should stay pretty constant over the life of the engine, though it will taper off a bit as the oil in the crankcase breaks down between oil changes. Keep track of oil pressure enough to note any sudden change in pressure patterns. If pressure drops off suddenly it's probably time to check the bottom end, if it's not too late.

If you're going gauge-happy, consider an accurate water temperature gauge to complement your boost and oil pressure gauge, since the stock water temp gauge is neither accurate nor particularly sensitive. Along with boost and oil pressure, water temperature will help you keep your engine happy.

Exhaust Temperature Gauge

A gauge that is commonly associated with turbo engines is the exhaust gas temperature (EGT) gauge. While some people like to use the EGT for tuning the engine's fuel mixture and ignition timing, it's best use is as a way to watch out for dangerously hot conditions in the exhaust stream. If the exhaust gasses get too hot (above 1,700 degrees F), it means the combustion temperatures are also very high and your engine could be suffering from a lean air/fuel

mixture or timing that is dangerously retarded. If you install an EGT gauge on your basically stock engine it will also give you a good baseline for later tuning. We'll get into this more in the chapters on EFI and tuning, but it's something to think about in the preparation stage. On the other hand, if you're just going to bolt on a few parts and not tune your car, or pay someone else to tune it, you probably don't need an EGT gauge.

A trick for installing an EGT probe is to center punch and drill your exhaust manifold most of the way through. Before the drill bit breaks through, start the engine and finish the hole to blow the chips out with positive exhaust pressure. This will prevent any chips of cast iron from passing through and possibly damaging the turbo and catalytic converter. The safest way to install an EGT probe is with the manifold on the bench, of course, since that will eliminate the possibility of contamination.

Once you've settled on a set of gauges to keep an eye on your motor's health and performance, you've got to mount them somewhere in the interior that is clearly in your line of sight. The cleanest installs make use of mounting kits that put the gauges in place of the center console vent (1g), radio or DIN pocket (Evo), steering column cover (Evo), or on the A-pillar (1g, 2g, and Evo). You can also buy universal cups to mount gauges wherever you

One of the most common gauge mounting locations is the driver's-side A-pillar. This puts the gauges right in the driver's line of sight without being too obtrusive. Aftermarket gauge mounting pods like this one make for a clean install that almost looks factory.

The center of the dash is a good spot to mount gauges if you have a VR-4, but other locations work well, too. At a minimum you should install a quality mechanical or electronic boost gauge. Add a good-quality EGT if you plan to tune your engine.

This paper-and-metal silencer found in the outlet of a 1g MAS is a small restriction. Remove it if you have the MAS out for another reason, but it probably won't increase power much on its own.

The 1g metal air filter housing is a serious restriction at higher-than-stock boost. Ditch as much of it as you can. The easiest solution is to remove as much of the outside wall of the housing as possible, for example by drilling large holes as this owner has done.

This one has no lid at all. The filter is an aftermarket oiled-cotton element, which is nice since it will last as long as the car. While this will result in warm air entering the turbo (colder air would be better) it frees up the intake for much higher power levels.

want. Pick one you like and spend a few hours installing it before tuning your car and you will have a much better idea of what's going on under the hood.

DSM vs. Evo Bolt-Ons

The 4G63t installed in the 1989–1999 Galant VR-4, 1g Eclipse/Laser/ Talon, and their 2g counterparts have some pretty significant differences compared to the 4G63t found in the USA-market Evo VII, VIII, and IX. Not only is the engine (and head) rotated compared to the early cars, the engine is also in quite a different (much higher) state of tune and features a totally different manifold and turbo arrangement. For that reason, we've broken them out for the purposes of this chapter.

Because they need more help, we'll talk about the early cars first. Yes, the 1g and 2g cars have some important differences that will impact the natural progression of bolt-ons, but we'll mention those as they come up. There are enough similarities, and enough interchangeable parts that they can be treated almost the same.

Once we've covered the DSMs, we'll talk a bit about the Evo, and how much power can be squeezed out of the engine without getting your hands dirty (or at least not completely dirty).

Air Filter or Air Intake

On the intake side of the engine (before the turbo) the 1g and 2g DSMs are different enough that there are two approaches to improving the flow here. In the years that have passed since the 1g intake tract was designed, engineers have figured out much more efficient ways of getting air into the engine, and it shows. The 1g intake and filter is a restriction at stock power levels, while the 2g intake and filter is a restriction only at slightly higher boost levels. The stock intake and filter on the USA-market Evo, in comparison, does not present as much restriction even for high power levels.

Start your intake mods on the 1g by removing the stock paper/metal silencer found in the tube leading from the filter to the MAF housing. This is a major restriction that is very

easy to eliminate. The second restriction in the 1g intake is the stock air filter housing. The old-school metal housing and small inlet are serious impediments to intake flow, and therefore to high-RPM and high-boost power.

The 2g intake is much less restrictive; you should be good for moderate power levels and boost with the stock setup. You can replace either filter with a drop-in reusable filter if you want, but it won't make much difference from a power standpoint. The stock filter has plenty of area for the engine to breathe through.

A reusable cotton-element filter will help when it comes time for regular maintenance. It won't have to be replaced every time you service the intake, and can be cleaned many times before it must be replaced. Make sure not to over-oil the filter, and go with a well-known brand like K&N. Excess oil on the cloth can migrate to the Mass Air Sensor (MAS) and cause it to read incorrectly.

For a slightly cleaner install, you can buy a generic clamp-on filter to

The stock 2g filter and housing is a good design, but a simple, inexpensive adapter and clamp-on air filter will allow more airflow to keep up with increased power. This should be able to handle anything you can do with a stock short block.

The 2g BPV is another weak spot. The stock valve can leak at stock boost levels, not to mention anything higher. Replace it with a 1g valve if you're working with a budget, or an aftermarket BPV like this one if you want something a little nicer.

mount to the MAS housing (an adapter is needed for the 2g). These are nice to have but by no means necessary—spend your money somewhere else first. Cold-air intakes, (CAIs), are so-called because they take air from outside the engine compartment, where it is generally cooler. The colder outside air is denser than hot engine-compartment air, and therefore the engine takes in more of

it. This can increase engine power, at least in theory. In practice, there aren't any true cold-air intakes for the DSM cars, so most are just a glorified MAS adapter and cone filter.

On the topic of intake improvements, there is one part of the 2g

intake system that should be replaced as soon as possible; the stock plastic bypass valve. The 1g comes with a really nice quality metal bypass valve (BPV) that will withstand as much as 18 psi of boost before it leaks. The 2g BPV, on the other hand, leaks boost at even stock boost levels (somewhere between 11 and 12 psi).

The easiest and cheapest way to replace the 2g BPV is either with an adapter that lets you use the 1g BPV, or with a 3000GT BPV. The second option is cheaper, but the first is more flexible, since many aftermarket intercooler pipes are designed with the 1g valve in mind. The 1g valve can be modified to hold even more boost pressure if needed.

As for the USA-market Evo, the intake system is very well developed. It is not as restrictive at stock power levels as the 1g and 2g intakes, but it does have its downfalls, particularly at very high tuning levels. In contrast

The biggest improvement on a 2g will come from a better MAS to turbo tube. The stock 2g tube has a massive restriction in the form of the BPV tube end. Removing it is actually such a common modification that many used 2gs have already had it removed.

Everybody wants to throw a blow-off valve (BOV) on his or her car for the look and sound, but it's really not a good idea unless you have a way of tuning the engine management system. A BOV on a stock car is a recipe for poor running, and it's not needed.

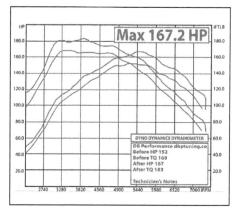

This dyno chart shows what happens when an otherwise stock 2g Talon gets a 3-inch exhaust system with aftermarket catalytic converter. With no other changes, the engine gains an impressive 15 hp and 15 ft-lbs of torque everywhere in the rev range. This is power you can feel. (Photo courtesy Travis Thompson/CarTech)

The Evo intake system is very good, but that doesn't mean it can't be improved. This very nice carbon-fiber cold air intake from Japanese tuner HKS Kansai is one of the best available. It draws in cold, high-pressure air from under the hood to make the turbo's job easier.

with the earlier cars, actual cold-air intakes are available for the Evo, including some that are very impressive in engineering and fit and finish.

Exhaust

The first thing most enthusiasts do when they get home with their new car is drop the exhaust system and install an aftermarket one. This isn't a bad idea once you've got the engine tuned up and running on high-octane fuel. The right DSM exhaust will pick up more than a few horsepower, and it can definitely improve the sound of your engine.

The most common (also cheapest and easiest to replace) exhaust systems are cat back types that replace the stock exhaust behind the catalytic converters. They're popular because federal law prohibits modification or removal of existing catalytic converters, making cat-back exhausts the default choice for most people.

It's been said before, but it's still true—a turbo engine wants as little exhaust as possible. Use the largest exhaust tubing you can squeeze in. A 3-inch exhaust all the way to the tailpipe adds a few HP and much quicker turbo spool-up.

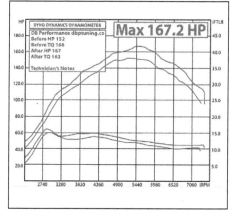

A closer look at the Talon's horsepower curve with the boost graph laid over it shows that the gains come from increased boost. From a peak of 15 psi to a high-RPM falloff of 10 psi, boost from the stock T25 increased to 16-psi peak and 11 psi at high RPM. This shows that even the tiny stock 2g turbo is flow-limited by the stock exhaust. There isn't much to gain even by increasing boost on such a small turbo; it won't be able to increase high-RPM boost. (Photo courtesy Travis Thompson/CarTech)

While you're thinking about and working on the exhaust, take a look at your exhaust manifold, particularly if you have a Galant VR-4 or 1g DSM. Check it for cracking and leaks. The cast iron of these early manifolds is very thin and is usually cracked. In fact, any non-cracked 1g exhaust manifold has probably been replaced in the past.

A leaky manifold looses exhaust pressure, which lowers boost and engine response, and can even cause damage to underhood wiring and hoses because of the heat. It might also kill you, too, since it is letting carbon monoxide into the engine compartment and possibly into the cabin of the car.

If you find cracks in your manifold, there are some things to watch out for when swapping on the new one. First of all, watch out for broken studs. The 1g and VR4 are notorious for having exhaust studs break off in the head. Buy a new set of studs before you start on your exhaust manifold work. Also get ready with a drill and screw extractors to remove any studs that break off below the gasket surface of the head. It's more common than you might think, and it's no fun to find you've got a broken stud on Sunday night when you have to get to work in the morning.

Removing a broken stud is so tough that you might consider removing the head and bringing it to a machine shop. Some areas have mobile machine shops that specialize in broken bolts, so try that before giving up on it and pulling the head. The turbo-to-manifold bolts (actually one stud and three bolts) are not nearly as tough as the manifold bolts, and you might get lucky with them. You can replace the lone stud with a bolt to make it easier to change manifolds without pulling the turbo out in the future.

More Boost

Yeah, we know this is the part of the chapter that you want to read the most, but it's really much smarter to wait until you understand more about engine management, fueling, and turbo efficiency before you blindly crank up the boost.

It depends on which car (and therefore, turbo) you have whether increased boost is even worth it. The 1g DSM has a decently sized TD05-14b turbo, and will produce good power without heating the air too much. The 2g, on the other hand, has a tiny little Garrett T25 turbo that really isn't much use above stock boost levels. An additional 2–3 psi is all that can be expected from the T25 turbo, also known as the "T-too small," for its diminutive size. The TD05h-16G turbos installed on the various Evos, on the other hand, can easily squeeze out another 5 or even 10 psi without breaking a sweat.

All of the 4G63t variants can handle 2–3 psi more boost from the stock turbo without a problem. There is also enough headroom in all three engine management systems to properly fuel a few more PSI of boost and the resulting airflow. It should be obvious that you need to have an accurate boost gauge before you start playing with boost tuning. Start small and test your changes slowly.

Before cranking up the boost, make sure that you're getting the expected stock boost out of your car. Test it in a high gear, like fourth or fifth, to give the turbo enough time to build full boost before the engine revs increase too far. Manual-transmission 1g owners should see between 8 and 12 psi on the stock turbo and intake system, while 2g owners should see 10 to 15 psi. The Evo is a different beast; stock boost has a peak around 19 psi that drops to around 16 psi by redline.

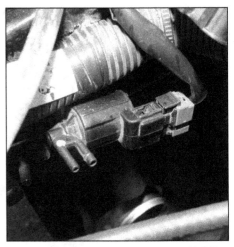

The 2g boost control solenoid (BCS) can be quickly modified for another 2.0 psi of boost by removing a restrictor on the output side of the solenoid (the larger port shown here). The restrictor prevents the boost signal from dropping quickly, which stops boost spikes in the stock boost control—it also reduces the stock boost level.

One thing to remember about the 1g and 2g boost control system is that it is not completely controlled by the ECU. The ECU can only reduce boost, not increase it. Removing the restrictor increases the target boost, as does adding an aftermarket boost controller. The second thing to remember is that the 1g and 2g DSMs have an upper limit to boost in the form of the factory "fuel cut," which kicks in if the stock ECU senses more than a preset amount of airflow into the engine. The fuel cut operates at around 18 psi depending on model, year, turbo, and air temperature. The only way to remove the fuel cut is with engine management modifications (see Chapter 6 for more information).

There are a couple of ways to increase boost levels, ranging from the cheap and simple to the complex and expensive. The simplest way is a bleed orifice that bypasses the BCS, an elaborate version of removing the

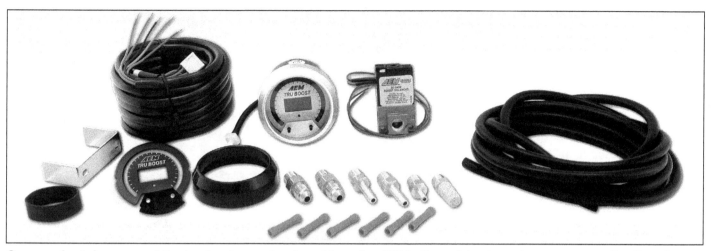

Once you've taken care of the basics, increasing boost by a few PSI is the quickest way to more power. Don't go in blindly, though, make sure you have all of the information you need including boost and exhaust gas temperature. A nice electronic boost controller, such as the AEM Tru-Boost unit, is one of the easiest ways to tune boost.

2g restrictor. To do it yourself, simply put a tee in the line between the turbo compressor and BCS, and another between the BCS and the intake tube, and connect the two tees with a length of vacuum hose.

Stick different sizes of restrictors in the hose (such as crimp connectors, vacuum restrictors from the auto parts store, or motorcycle carburetor jets) until you get the boost level you want. Tuning such a system requires swapping a lot of restrictors, but start with a small one and work your way up.

A slightly more complicated version of the same thing is a mechanical boost bleed valve with a ball valve or variable orifice (the so-called Grainger valve—Google it). These are nicer than a simple boost jet, and are good if you want to have more than one boost setting, or want to play with your boost pressure without swapping parts. A good MBC (not the eBay specials) will be very consistent and very quick to tune.

The most precise way to control boost (without engine control modifications) is with an electronic boost controller (EBC). These can be expensive, but some people prefer the pro-

grammability and increased response they provide over a mechanical valve. EBCs reduce turbo lag by quickly responding to a precise pressure point by opening the wastegate actuator.

No matter what method of boost control you choose, it can be hard to tune. At a minimum, you will have to

perform a lot of high-gear pulls to establish a good boost level without spikes or excessive lag. Try to set your valve or controller so that it opens a little before your target boost and allows the wastegate to prevent uncontrolled boost, and "creep up" on your target boost a little at a time.

If you've got a chippable DSM ECU, a simple reprogrammed chip, and some electronics experience, it's not hard to add a couple of horsepower to your engine. An off-the-shelf tuned ECU chip will provide some benefit but not as much as a custom-tuned program. See Chapter 6 for more about how to do your own tuning, or what's involved when you pay someone else to do it.

There is a maximum boost that you can run with your combination of engine, turbo, and fuel. Any more boost will cause your exhaust gas temperature (EGT) to get up into dangerous territory, or cause detonation or turbocharger failure. That's why it is always better to slowly tune up the boost. Set it at 10 psi for a while until all of your fuel and timing curves look good, and then slowly increase it with each run. Watch your EGTs and check carefully for knock at each increase in boost.

Tuning or Chipping

One of the wonders of modern EFI systems is the control they give over the engine, both to the original engineers and to the aftermarket. The hard part, for the tuner anyway, is taking advantage of that control. The stock ECU is set up to cope with low-octane gas, maintenance-neglecting owners, and warranties—three things that the enthusiast owner won't have to worry about.

At this stage of modifications (2–3 psi more boost, free-flowing exhaust and intake) a modified ECU will provide some additional gains. Your engine will probably be able to handle more ignition timing and more RPM than stock. It might also be running richer or leaner than is necessary or safe. Tuning all of these factors can give an enormous power boost to an engine that is outside the bounds of the stock ECU.

Unfortunately for DSM owners, only some ECUs can be tuned. Some of them do have the calibration program burned into a chip that can be removed. These so-called EPROM ECUs can be modified by a tuner with a replacement chip that increases timing and allows the ECU to cope with more boost and higher-rpm running. See Chapter 6 for more details on

On a pure "dollars per horsepower" scale, a turbo timer ranks pretty close to the bottom. You won't make any more power, but it's easy to install. Plug-and-play harnesses are available for DSMs and Evos, so it's not a bad installation for practicing your electronics skills.

chippable ECUs. The EVO can be flashed with a special interface cable and a laptop computer, which opens up much more room for tuning.

It's been said before, but it bears repeating: A few hours spent on a dyno with a tuned stock engine are worth every penny. Rather than trying to find the right aftermarket chip or tune to work with your parts, you will have a program that works with your engine exactly the way it is. All of these topics are covered more in-depth in the next chapter.

Other Parts

There are some other bolt-on modifications that show up in the pages of our favorite import magazines or are hotly debated on Internet car forums. Some of the parts work, and some don't. The most popular (after discarding the truly worthless, like battery-powered intake blowers) are turbo timers, so-called grounding kits, and so-called underdrive pulleys. Since we don't discuss much about these parts later in the book, this is as good a place as any to detail the pros and cons of each.

Turbo Timers

A common bolt-on product is a turbo timer. Basically this is a small device that keeps the engine running for a preset time after you turn the ignition key off. This keeps the oil and water flowing to the turbo as the turbine blades spin down. This helps the turbo last longer.

A turbo timer won't help your engine make any more power, but it is helpful if you like to jump out of your car as soon as you park it. Otherwise just wait a few seconds while the turbo spools down before shutting off the engine.

Grounding Kits

Within the last three or four years, so-called grounding kits have become very popular as a bolt-on engine modification. The theory behind these kits is that the electrical connections between the engine and chassis may not be good enough to transmit all of the low-voltage signals from various sensors to the ECU, and the high-current signals from the ECU to the injectors and coils.

In practice, these kits vary widely in effectiveness, and are probably not worth the money. If you want to improve your ground connections, build your own ground straps with crimp terminals and large-gauge wire. On older cars with more corrosion between the terminals, they may be more effective. Even there, it would probably be a better use of your money to do some of the other modifications listed in this chapter.

Underdrive Pulleys

An engine's accessories, including the A/C compressor, water pump, alternator, and steering pump absorb engine power as they are driven by the accessory belt system. Reducing the diameter of the crankshaft's drive pulley reduces the speed that these accessories are driven, which in turn, reduces the power demands they make on the engine.

So-called underdrive pulleys do just that. The smaller, usually aluminum pulley also helps by reducing

In theory, lightweight aluminum pulleys are a good thing, because rotating mass absorbs horsepower as the engine revs. The less rotating mass, the faster the engine will rev, all else being equal, and the more power can be used to accelerate the car. But in practice they aren't that impressive.

the engine's rotating mass. There are other pulleys on the market that are lightweight aluminum, like underdrive pulleys, but they are designed to drive the accessories at stock speeds. These pulleys may have some incremental effect on the engine's rotating mass, the same as an underdrive pulley.

However, underdrive pulleys are not usually a good idea on a street-driven DSM or Evo. They can cause trouble with charging and cooling if they are small enough to reduce accessory power demands, for one thing. For another, the stock crank pulley contains a rubber ring designed to damp torsional vibrations that can damage the crankshaft or bearings. No aftermarket lightweight pulley features such a damper, meaning that they might have an unexpected side effect of shortening crankshaft life.

On a race car, lightweight pulleys are a good idea. Usually the engine is not run long enough between rebuilds to worry about bearing and crankshaft life, and the stock accessories could stand to be run at a lower speed if the engine sees a lot of high-RPM use. Water pumps, in particular, act strangely if they are driven at too high a speed. A street-driven car should stick with a good-condition stock pulley whenever possible (check yours for cracking and separation of the rubber ring, incidentally).

Motor Mount Inserts

In FWD and AWD cars, the engine and transmission absorb all the drive forces that act through the front tires and axles on the car's chassis. When the driver dumps the clutch, the shock loads rock the engine and transaxle back and forth violently and can cause the car to become unsettled and lose traction. A less violent version of the same

Hard urethane inserts for the factory DSM (and Galant) motor mounts will help traction and shifting. They're not very expensive or hard to install, but they do transmit a lot of extra vibrations to the body shell and make the car noisier.

thing happens during shifts; the engine and transmission move back and forth, making shifting difficult.

The factory DSM engine mounts make the problem worse, since they are soft in order to isolate vibration from the passenger compartment. Luckily, several aftermarket companies make hard, yet flexible, urethane inserts that fill the voids in factory motor mounts. While this is not an engine modification, the resulting decrease in movement helps the handling and shifting of a car, which can result in faster drag times and more positive shift feel.

You can also make your own filled mounts with room-temperature curing urethane. Sold in auto parts stores for gluing windows into cars, 3M's window-weld adhesive is the most commonly available. Use it to fill in the voids in the factory soft motor mounts to stop them from moving as much as stock mounts.

TURBOCHARGING AND INTERCOOLING

Turbochargers and intercoolers are the key to the 4G36t engine's performance, both stock and modified, but they need to be looked at and modified in the context of air flow, fluid dynamics, and gas theory.

You don't need to know everything about airflow to pick the right turbo for your car, but it will certainly help when you call vendors and do your research before buying. It will also help you to manage your expectations of what power your engine will be able to put down, and what a particular turbo will be like to live with on a daily basis.

Volumetric Efficiency

There is a way to measure how good an engine is at getting air in and out: Volumetric Efficiency (VE). VE measures how well the engine takes in and expels air through the intake, cylinders, and exhaust. Basically, VE is a measure of engine airflow efficiency. Even though an engine might measure 2,000 cc in mechanical displacement (bore area X stroke), it will probably not take in that much air when running. The VE is the volume of air taken in with each cycle divided by the total displacement of the engine. It is normally expressed as a percent (of displacement).

Engines are unable to take in enough air to fill their displacement for a number of reasons, including friction between the air and ports, restrictions in the intake and exhaust tracts, inertia, and valve timing. The turbo adds another layer of confusion to this story, but VE works the same for turbocharged and non-turbocharged engines, though the numbers may be different.

VE is not a static number. As an engine runs, the VE varies according to RPM and load (a measure of how hard the engine is working; load increases with greater throttle openings but does not match throttle position exactly). VE is closely related to engine power, although it is not the only factor influencing output. Assuming that everything else is the same, an engine with a greater VE will produce more power than an engine with a lower VE.

One thing to keep in mind about VE is that an engine's VE has nothing to do with part-throttle power. The throttle body is the primary restriction in a gasoline engine, so if the throttle is not fully open, the engine is not producing anywhere near its maximum VE. At part throttle, the only way to get more power is to open the throttle more.

That is why a racing engine can be designed for maximum power and VE at all costs—the time spent with the throttle closed is minimal. A street engine, on the other hand, spends most of its time with the throttle cracked part way open, so a wide, flexible powerband is more important than ultimate efficiency.

Incidentally, engine airflow, in the USA at least, is measured in cubic feet per minute (CFM). There are a couple of other measures of engine efficiency that will be mentioned in more detail later in the intake and engine management chapters, but VE is the most important one. Almost every engine modification that we perform is done with the idea that it will improve engine VE, and thus performance.

Enter the Turbo

The volume of air allowed into the engine is actually only a convenient indicator of the number of air molecules getting into an engine.

The actual molecules in a given volume of air combine with fuel in the combustion chamber to make power. The more molecules in a given volume of air, the more that air weighs, and the greater its mass. Since the volume of airflow through the engine is limited by VE, engine displacement, and RPM, the only way to get more air into the engine is through a greater mass (weight) of air in the same volume.

It helps to understand the nature of gasses, like oxygen and air, to make sense of how a turbo works. The number of molecules contained in a given volume of gas (and thus the power potential in reference to shoving it in an engine) is determined by the pressure and temperature of the gas. This relationship is captured in the Ideal Gas Law, one of the principles of physics. The whole thing is rather complicated but in a nutshell: As you decrease temperature, the number of molecules increases; as you increase pressure, the number of molecules increases.

Given a known volume of air, a known temperature and a known pressure (in bar), you can calculate the number of molecules (and thus the mass) of that air. You cannot change one factor of the equation (pressure, volume, temperature, or mass) without affecting the other factors.

The mass of a given volume of air is sometimes referred to as air density. In any case, an engine will make more power at the same RPM in denser air vs. the same engine in less dense air, such as on a hot day or in the mountains. Think of it this way: At sea level, the engine's intake "sees" 14.7 psi of air pressure, otherwise expressed as 1 bar of pressure. By doubling that pressure to 29.4 psi without increasing the temperature of the air, you have theoretically doubled the number of molecules that

A turbocharger compresses the air coming into the engine so that more air molecules fit into the same volume of air. Each engine stroke takes in more air molecules, which can be mixed with more fuel molecules to produce more power.

The compressor (on the left) is driven by a turbine (on the right), in the exhaust stream from the engine. In a way, the turbo is driven by "wasted" energy in the form of pressure and heat in the exhaust.

the engine takes in with each revolution. This holds only as long as the other factors stay the same, meaning the same temperature, and the same volume of air.

That means that in a perfect world, you can mix the air with more fuel, and you have just doubled the potential horsepower of that engine. Of course in the real world things are not that simple, but the idea still holds—more pressure is more air molecules is more density, which means more power as long as temperature, volume, and engine VE stay constant.

Turbo Theory

So the basic idea is more pressure = more density = more power. While it might seem like magic, the turbo manages this amazing feat simply by pumping denser air into the intake manifold than would otherwise be available.

A turbocharger is a kind of air pump known as a centrifugal compressor. Basically, air is sucked into

the center of a spinning wheel and then flung outward at a high speed. A combination of kinetic energy and speed causes the air molecules to get closer together, increasing the density at the outlet of the turbo, and therefore into the intake manifold.

Without the turbo in place, the engine would breathe easier through the exhaust, but the compressed air flowing into the intake manifold more than makes up for the additional backpressure caused by the turbo. With a high-powered turbo engine it helps to think of the engine as a giant exhaust generator for the turbo rather than the turbo as an air compressor for the engine. The turbo determines the airflow patterns of the whole system, even though the engine does the work that makes it to the axles. This has huge effects on the power curve of the engine.

Unfortunately for performance enthusiasts, the turbo heats the air as it compresses it. This is the result of friction and the Ideal Gas Law—the turbo increases the pressure, which forces the temperature to increase along with density. The air coming out of the compressor is therefore hotter than the ambient air going in. Given the above note about pressure, you would think that this reduces the effectiveness of the turbocharger, and you would be right. The hotter the air, the less dense it is; the less dense the air, the fewer oxygen molecules find their way into the engine resulting in less fuel burned and less power made by the engine. However, as with the exhaust back pressure caused by the turbo, the increased air temps reduce, but do not eliminate, the benefits of feeding more dense air to the engine.

Compressor Efficiency

At a particular combination of airflow and pressure, a turbo's compressor heats up the air more than expected because it is not 100 percent efficient. The difference between actual and expected discharge (output) temperature is determined by the turbo's efficiency at a particular combination of airflow and pressure. Turbo efficiency varies significantly between different turbos and different combinations of pressure and airflow. All else being equal, a more efficient turbo heats the air less, creating a denser intake charge, and therefore making more power.

That's why in some cases a larger turbo can make more power than a smaller turbo at the same boost level. The larger turbo running 15 psi at the same CFM is actually in the sweet spot of its efficiency and able to blow that air much cooler than the smaller turbo operating outside its efficiency range. You can find turbo efficiency on its compressor map. The compressor map is a chart showing the efficiency at various boost and airflow points, therefore allowing you to find the

Size is not all that matters. Compressor blade design is critical to turbo efficiency and performance. This compressor, from a standard-rotation Evo III 16G turbo, has fins that vary in thickness as well as height, for the best combination of airflow, compressor flow range, and spool-up speed.

right turbo for your desired boost and airflow combination.

The pressure that matters to the turbo for this purpose is not gauge pressure or pressure above ambient; it's the pressure *ratio*, or relationship, between the turbo's input and output pressure. For example, at sea level at 68 degrees (standard pressure), ambient air pressure is about 14.7 psi. If the engine is running in such conditions with a manifold gauge pressure of 14.7 psi, this figure is above the 14.7 psi ambient, so the pressure ratio is 2:1 (29.4:14.7).

If you know the airflow of your engine and how much boost you're making, you can use a compressor map to figure out the efficiency of a particular turbo at a particular RPM. It requires plotting a few boost and airflow points on the map, and reading off the efficiency.

When you're choosing a turbo there is such a thing as a too big or too small compressor (or "cold side"). If you pick a compressor that's too big

Some aftermarket turbos offer advantages you'll never get from a bolt-on turbo, like these surge ports on the compressor inlet. These expand the turbo's effective airflow range, and make it more efficient during spool-up than a similar turbo without surge ports.

for your engine you might run into a situation called surge. Compressor surge is essentially flow instability and can result in flow reversal at some locations on the wheel inlet. This will damage the turbo and drive intake temperatures up, as compressor efficiency in this region is very poor.

On the other hand, a compressor that is too small will choke off airflow to the engine at high engine speeds since the engine requires more airflow at high RPM. The turbo won't be able to keep up with the engine's airflow demands above a particular boost level. When this happens, the turbo has reached its boost ceiling, and boost will begin to drop as engine RPM increases. You can bet the turbo is not at its most efficient near the boost ceiling (such as a T25 at 16 psi).

Do you need to start your turbo search from scratch and read the compressor map of every turbo on the market? Not really. The DSM and Evo motors are very well known these days, and there are a lot of good turbo choices that have been proven by dyno, drag strip, road race, and autocross performance. Talk to your turbo supplier about your goals and budget and they'll help you pick a turbo that will work best with your combination. You can learn a lot about your engine's airflow demands and turbo behavior by going through the turbo selection process, so we've given an example in the sidebar on page 42.

Exhaust velocity in a split-scroll setup is also kept high by the small volume of the split exhaust manifold compared to a traditional 4:1 manifold. Twin-scroll turbos and manifilds are very good for a street or rally car, but the turbo selection is better for single-scroll setups.

Turbine Efficiency

The exhaust turbine determines much of the boost characteristics of the turbo. A smaller exhaust scroll with a small, light wheel will accelerate the compressor wheel faster and bring on boost sooner than a larger slower wheel. However, at high engine RPM and load, when the volume of exhaust gas being created becomes very large, the smaller turbine will become a restriction. It will not be able to get the gas in the exhaust manifold out fast enough, increasing exhaust manifold pressure.

Exhaust manifold pressure limits power production because the engine can't get exhaust out of the cylinder well enough on each exhaust stroke. Once exhaust pressure reaches or exceeds intake pressure, the turbo isn't able to increase airflow through the engine very much and the turbo has become a restrictor instead of a helper.

Along with turbine wheel size, the geometry of the turbine housing is an important factor for determining how fast a turbine spins up to operating speed. Since the scroll diameter is based on the turbine wheel diameter, the important number is the ratio between the area of the turbine input and the radius from the center of the turbine inlet to the center of the turbine wheel. This is known as the A/R ratio for Area:Radius. If the A/R is very small, the gasses hit the turbine very quickly and get it moving with less delay. However, this means that the area of the exhaust inlet is small in comparison to the wheel diameter, and will limit maximum gas flow rates.

Not all turbo manufacturers publish the exhaust A/R ratio, although most publish the area of the exhaust inlet. Assuming the same wheel, the greater this area, the greater the A/R ratio.

Unfortunately, the perfect exhaust turbine doesn't exist. The higher the exhaust flow capacity a turbine, the slower it spools, and vice versa. The flip side of this relationship is that sometimes you will get the same power or less if you increase the boost on a turbo that's too small on the exhaust side. Raising the pressure ratio at a particular flow rate could bump the turbine out of its efficiency range. Instead of producing more power, you're simply choking off engine exhaust flow in order to heat up the intake air—not a good tradeoff. Power will start to fall off, and you'll find that you are reducing ignition timing to avoid knock with increased boost (but no increase in power).

Many modern turbos, like those installed on the Evo VIII and IX, are very close to the perfect turbo for increased spool without a large reduction in top end flow. These turbos feature an improved exhaust housing that uses a twin-scroll design. The idea is to separate the pulses from each pair of cylinders so that they reach the turbo independently.

Since each of the two turbine inlets in a split-scroll setup is smaller than one large inlet, the effective A/R ratio of the housing is lowered. This improves spool-up and lowers the boost threshold compared to a similar non-split turbine, since the energy of each exhaust pulse is not lost to negative pressure from another cylinder's exhaust-port opening.

Reading a Compressor Map

It's not usually necessary to read a compressor map and choose a turbo if you're building a proven engine with a huge aftermarket like the 4G36t, but it's a good way to get a feel for the kind of performance you can get from a particular turbo. Stepping through the process at least once will also help you understand how airflow and boost interact in a particular turbo/engine combination. It can also help you tune your engine by showing you the most efficient boost point. The ability to read a map can also serve as your BS-detector when bench racing with friends. You'd be surprised at how many people have the wrong turbo for their application or aren't taking advantage of the expensive turbo they have.

To simplify our example and make the whole process easier, we'll assume sea level pressure (14.7 psi) and standard temperature (68 degrees F), just like the maps are scaled. We're also going to ignore some of the effects of the Ideal Gas Law (the heating and efficiency loss as the air is compressed). We'll also set aside the effects of pressure loss between the turbo and intake manifold.

Yes, this makes our example a major over-simplification of airflow and VE under boost, but it should help you make sense of turbo compressor maps and the basic principles involved. If you want more in-depth information on airflow and more accurate calculations, there are some excellent turbo calculators on the Internet.

The first step is finding a compressor map. This sounds easy but it's not. Not all manufacturers make theirs available to anyone who wants it. MHI is particularly bad in this respect. They don't generally release their maps to the public. That's why the attached map is not a map of a MHI turbo known as the "Evo VIII 16G6." Instead, it's a map of a turbo called the "HMI OVE IIIV g16," which happens to flow just like a MHI turbo. We have re-scaled it to read in CFM although MHI's is rating mass flow in kg/cm³.

Once you have a turbo flow map, blow it up on your computer or with a copier until it fills a full 8.5 x 11-inch piece of paper. This will help you plot points on the map. We're going to plot three points on the map to show how airflow and boost change through the engine's operating range. For each point we need to figure the engine's airflow and potential boost. With those two numbers, we can find the compressor airflow (since the compressor map is denoted in compressor inlet mass airflow rather than outlet airflow, or volumetric airflow).

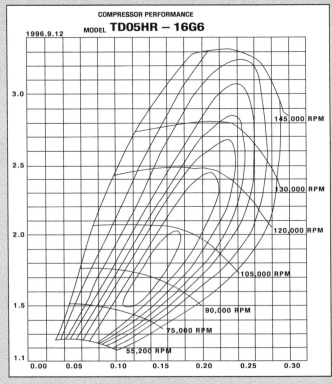

The three points are going to be right at the onset of boost, or the minimum RPM where you might expect to go full throttle; in this case, say 2,500 rpm. The second point will be at peak torque; for this we'll assume it's a stock Evo VIII, with peak torque around 3,500 rpm. Finally, we'll plot the airflow up near redline. The Evo VIII 4G36t has a redline of 7,600 rpm.

Each point will use full-throttle potential airflow, determined by the displacement of the engine and current VE. Assuming a stock 85-mm bore and 88-mm stroke, the displacement of the engine is 1,997 cc. Multiply by .061 to convert to cubic inches (121.82 ci in this case). Divide by 2.0 to get maximum potential airflow per revolution (since it's a four-stroke engine, each cylinder only fires every other rotation), resulting in 60.91 cubic inches. Since we'll need this number in cubic feet for the compressor map, divide by 1,728 (the number of cubic inches in a cubic foot), resulting in .035 cubic feet of potential airflow per revolution.

Now we calculate the airflow at our first point, 2,500 rpm. The engine has not come up "on the cam" and is not very efficient at this RPM. It also isn't producing much boost. We'll assume that VE is around 80 percent to make things easy.

Reading a Compressor Map CONTINUED

Incidentally we always use max VE, because that is what you get when the throttle is wide open, and that's the only way to guarantee maximum possible boost at a given RPM.

Take 80 percent times our potential airflow of .035 cubic feet to get .028 cubic feet of air per revolution. At 2,500 rpm, that means the engine is able to take in 71 cubic feet per minute. At 2,500 rpm the engine is unlikely to be boosting at all, so say 1.0 psi at most. At our standard temperature and pressure, this is a pressure ratio of 1.07 [(14.7+1)/14.7]. That means that the compressor is sucking in 75 cfm (71 X 1.07). Plotting this point on our graph (1.07 pressure ratio, 75.44 cfm) puts us in the bottom left corner of the turbo map, basically off the chart. That's fine, because there isn't any power to be made at this RPM. It's just a good starting point for our engine's boost curve.

Now let's look at our second point. At 3,500 rpm, the engine is running more efficiently. VE is very high at this point. On the well-developed Evo VIII engine, VE is probably better than 95 percent at torque peak. The stock boost control peaks boost at roughly 20 psi at this RPM. That's a pressure ratio of 2.36 at sea level [(14.7+20)/14.7]. Engine airflow at this RPM is 116 cfm (.035 X .95 X 3,500), which means that compressor airflow is roughly 274 cfm (116 X 2.36). Plotting this point shows that we're in the 68–70-percent efficient range, a little outside the optimum.

The third point, 7,600 rpm, is the absolute maximum RPM point. VE has declined from its peak by this point, but is still high, probably in the 90-percent range. The engine's VE here depends heavily on the cams. A stock-cammed engine will be falling off more than one with big aftermarket cams, and the Evo will be breathing much better than a DSM 4G36t with its smaller cams. The stock boost control tapers off to around 16 psi at this RPM. Going back to our calculation, we have 239 cfm of engine airflow (.035 X .90 X 7,600), and a pressure ratio of 2.08 [(14.7+16)/14.7]. Compressor airflow is an amazing 499 cfm, at 71-percent efficiency.

From this exercise, you can see how well the stock Evo VIII turbo is matched to the engine. The efficiency stays fairly high at all airflow points, and none of the boost points lies outside the critical "surge line" of the turbo.

Now let's look at what happens when we add an MBC to our Evo and eliminate the factory boost taper. Instead of 16 psi at redline, now we're able to run 19 psi. The pressure ratio is 2.29 instead of 2.08. With the same engine airflow, 239 cfm, we now have 547 cfm of compressor airflow. This pushes us out to 68-percent efficiency, which still pretty good for a stock turbo running nearly 20 psi at redline! Sometimes it's helpful to plot a few different boost curves on one compressor map to help you see what different changes will do for the engine.

Turbo Lubrication and Cooling

Turbo bearings often have to withstand over 100,000 rpm, so special plain or ball bearings are designed to work with those speeds, vibration (caused by minute imbalances in the wheels and boost-created forces), and high temperatures. They require a small, but steady flow of very clean oil, and they need a large, direct drain. The seals between the bearings and the turbine/compressor housings are not designed to seal, but rather minimize leakage by not being the path of least resistance, so oil has to be cleared out as quickly

A bad turbo oil seal pumps oil into the exhaust housing, fouling it up with sticky black deposits. You don't have to pull off the turbine cover to see the deposits, but it sure helps.

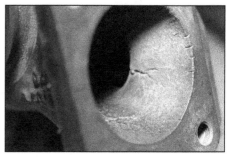

Another common turbo failure is cracks in the exhaust housing. If you're looking for a used turbo, be sure to inspect the area between the wastegate port and the turbine outlet. The casting gets thin here, and it's the most likely place to crack. Don't thin it out too much either, if you're porting the turbo yourself.

Save the "dry" (non-water cooled) turbos like this one for the track. While you could use it on the street, the lack of cooling will eventually cause the oil remaining in the bearings to harden and plug up, choking off the supply. From there it's just a short time until the bearings seize and gall from the lack of fresh oil.

There are several things to notice about this turbo installation, including the neat cast iron exhaust manifold and O_2 housing with provisions for an external wastegate. A ball-bearing turbo (like the massive one shown here) should always have a filter installed in the oil feed line to protect the ball bearings.

as possible. That's why the oil drain is much larger than the oil feed line. Keep your turbo fed a diet of clean, cool oil, and your turbo will last almost forever. Neglect oil changes on a turbo engine, however, and you'll be buying a new turbo.

All of the stock DSM turbos use water cooling for the bearings. Coolant prevents the oil trapped in the bearings when the engine is shut off from burning and turning to semi-solid. Water-cooled turbos are preferred on street cars and most race cars. Unfortunately some of the very large drag race turbos don't have water-cooled center sections, but in every other case it pays to hook up the water feed lines no matter how much of a pain it can be. If you're using a turbo designed for water cooling without the water, it will fail in every case.

Ball- vs. Sleeve-Bearing Turbos

Until recently, most turbos (and all stock MHI turbos) have plain sleeve bearings in the center section.

These bearings are tough and easily withstand the load and speed of a turbo shaft. They last a long time, because the oil between the two concentric bushings prevents direct contact between the various parts.

Turbo manufactures have begun adding ball bearings to their center sections. Ball bearings have the advantage of reduced friction, which allows ball bearings turbos spin up faster and create boost at a lower engine RPM than plain bearing turbos. Also, once the turbo is spinning, it reacts to changes in airflow much more quickly. Transient boost response (i.e., letting off and then accelerating again at high RPM) is vastly better than a plain bearing turbo with the same wheels and housings.

Of course, there are downsides to ball-bearing turbos. First, they are more expensive than the equivalent plain bearing turbo. In the aftermarket, the difference is as much as $500 or so. Second, the bearings inside the turbo are much less tolerant of sloppy assembly. They last longer than plain bearings, but the shafts must be precisely balanced and

aligned for a ball-bearing turbo to work well.

Ball bearings are also much less tolerant of dirt in the oil supply. While it's always a good idea to add an in-line filter to an oil feed line, it's a necessity with a ball-bearing turbo. Neglect the ball bearings and you'll be buying another rather expensive center section, but treat them with respect and you'll be rewarded with a turbo that spools quicker than a plain bearing turbo. This is particularly useful for autocrossing and road racing, where ball-bearing turbos excel.

Wastegates

If you just bolted a turbo to your engine, the exhaust turbine would spin freely, driving the compressor to

All stock 4G36t turbos have an internal wastegate controlled by this flapper door. The wastegate exits into a separate port in the exhaust turbine housing, from where it must be merged into the rest of the exhaust system. It's a good, reliable system that Is preferred for most street cars.

Notice how the Evo's turbo maintains separate flow for the wastegate valve from the manifold all the way out to the O₂ housing. This keeps the gasses from impinging on the flow out of the turbine, increasing efficiency and flow. You're much less likely to get boost creep with the Evo turbine than the DSM turbine.

There are a lot of reasons you might want to consider an external wastegate if you're building a high-horsepower DSM or Evo. For one thing, the wastegate valve is much larger than the stock internal wastegate, meaning boost pressure can be held closer to the set point with less wastegate operation. For another, the wastegate outlet doesn't affect the flow out of (and therefore through) the exhaust turbine.

Careful porting can help you avoid boost creep in some cases. You'll need a very steady hand to pull this off, though, since a slip could cause the wastegate valve to leak exhaust and slow spool. This particular turbo has a larger flapper valve installed as well, which is even trickier than porting the wastegate opening.

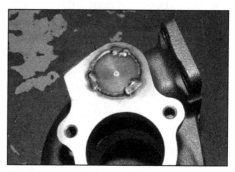

You can use the stock turbine housing easily with an external wastegate; just block the stock wastegate port on the outlet side. The cleanest solution is to weld a flat disk over the wastegate area, but you could also just wire the stock wastegate flapper closed.

a speed that would cause it to come apart. In addition, boost would increase beyond the engine's ability to physically handle it. To prevent this from happening, and give some control over boost, a wastegate bypasses some of the exhaust flow around the turbine. The wastegate may be mounted inside the turbine housing (as with the great majority of stock street turbos) or outside the turbo.

Most street cars will be fine with an internal wastegate. The stock wastegate inside most MHI turbos is decent, but sometimes it doesn't flow enough exhaust. If you're targeting moderate boost levels with a smallish turbine housing, you might find that an internal wastegate won't allow you to control the boost. What happens is the flow through the turbine is still so high compared to the wastegate flow that the turbo continues to build boost. This is called boost creep and it can be extremely damaging.

One way to eliminate boost creep is to increase wastegate flow com-

pared to turbine flow. Restricting turbine flow works pretty well—install a catalytic converter to do that and save the environment, too. This will naturally reduce power output, so most tuners prefer to increase wastegate flow instead. You can do this by porting the wastegate opening if the problem is minor, or replacing the

wastegate flapper valve with a larger one. If that doesn't work, you will have to go to an external wastegate.

External wastegates are easier to control, since they require less movement for the same amount of exhaust bypass. They can also increase power

This outlet elbow is another way to block the stock wastegate outlet for an external gate. Notice that it has the stock bolt pattern but no separate wastegate channel.

simply by eliminating the turbulence caused by an internal wastegate's flow hitting the main stream of gasses coming out of the turbine.

The wastegate is opened by a wastegate actuator containing a rubber diaphragm connected to a boost source. As boost increases, it pushes on the diaphragm, which in turn opens the wastegate and diverts exhaust around the turbo. The wider the wastegate opens, the more exhaust is "wasted" out the downpipe.

If the actuator is weak, maximum boost will be limited since the flapper valve can be blown open by large volumes of exhaust. This problem is known as wastegate "flutter." It can be solved in most cases by turning up the boost or changing the wastegate to one with a stronger spring.

Wastegate actuators are rated in PSI of boost. At the rated PSI, they open completely and prevent more boost from building. A 10-psi actuator will allow only that much boost unless a boost controller is used. An internal wastegate's actuator and its spring also have to resist the force of exhaust gasses hitting the flapper valve.

Better Boost Control

There are several ways to control the wastegate signal, including bleeding off some of the signal (like the stock boost control does). Directly connecting the wastegate to the intake and using a stronger wastegate spring is another way to control boost. Boost will only rise as high as the wastegate actuator permits.

The biggest downside to these kinds of boost control (increasing actuator strength or bleeding the boost signal) is that it slows the speed at which the turbo spools. Because the wastegate actuator is connected to a boost source even during spool-up, it is always being pushed open slightly. This bleeds a little exhaust around the turbine, exhaust that could be helping the turbine to spin up faster.

To get around this problem requires an interrupt-style MBC or an electronic boost controller (EBC). Both place a restriction in the line between the signal and the wastegate actuator. The hardest part of tuning an interrupt-style boost control system is adjusting the set point of the interrupter. In interrupt mode, both MBCs and EBCs have to be set to open a little before target boost is reached to give the wastegate time to open and bleed off exhaust flow.

The best way to control boost is with an aftermarket ECU with a MAP sensor that is able to do closed-loop

The easiest and cheapest way to increase boost is a simple manual boost controller. As soon as boost reaches a preset point, the valve opens and the boost signal hits the wastegate. This prevents any leakage of exhaust flow around the turbine and increases spool dramatically.

boost control. It may take longer to tune, but ECU-controlled boost is easier to integrate with the rest of your engine management system and tuning parameters.

Intercooling Theory

Without any way to get around the heat problem, a turbocharged engine would never live up to its full potential for power. Compressing air heats it up. This is bad, because the higher the air temperature, the more it expands and the fewer oxygen molecules are in a given volume (lower density). The heat also reduces the engine's ability to run ignition timing, which increases the tendency for detonation and engine damage.

What is needed is some way of cooling down the intake charge so that more of it fits into the intake manifold and more of the benefit of pressurizing the air is realized. Luckily we have intercoolers to slough off some of that heat by passing the hot compressed air through a cooling matrix. The air coming out of an intercooler is cooler, of course, and it's denser.

The intercooler matrix can be cooled by gas or liquid, but all production Mitsubishis use air-to-air intercoolers because they are lighter, less maintenance-intensive, easier to manufacture, and lower in cost. Most aftermarket intercoolers are also air-to-air, for the same reasons. Production air-to-water intercoolers are very rare. They're most often used in drag racing and land speed racing, because the runs are short and there's enough time to fill the tank with fresh, cooled water before hitting the track. A few people have built their own air-to-water setups for DSMs.

Intercoolers, like turbos, vary in efficiency. There are a couple of different ways of looking at intercooler efficiency, including heat transfer ability, heat capacity, and pressure drop. Since the main purpose of an intercooler is to reduce the temperature of the intake air, we'll look at heat transfer first. An intercooler that's 100-percent efficient will cool the charge air back down to the temperature of the cooling medium, which is ambient air for an air-to-air cooler.

The size and construction of the intercooler is also important for heat transfer. A larger cooler will do a better job of cooling the charge than the stock one. Also important is the temperature difference. On a cold day there is a larger temperature difference between the ambient air and the boosted air, which is the driving force of heat transfer. The temperature difference can be increased with an intercooler water sprayer (see below) or better airflow through the intercooler core. A larger front-mount aftermarket intercooler will have better airflow and therefore better heat transfer than the stock side-mount.

In the real world, it's harder to quantify intercooler performance than turbo performance. That's because on the street, intercoolers

All Evos from the very first Lancer Evolution have been equipped with a large, race-friendly, front-mount intercooler. This gives it much better access to cooling airflow than any of the stock DSM intercoolers.

don't actually work the way described above. They are not simply a kind of radiator that takes heat out of the compressed air supply and transfers that heat into the air passing through it.

About 90 percent of the daily driving we do is at part throttle with no boost, with short bursts of 90 percent or more throttle and high boost. During the time that we're not flogging the motor to make boost (and heat the intake air), the intercooler is just sitting there at basically ambient temperature. When we do crank up the boost, we quickly blast the intercooler with hot compressed air from the turbo. The hot air raises the temperature of the intercooler, which then passes that heat more slowly into the surrounding air, as well as the air going into the engine when we're off boost. Once we're done proving our point, the intercooler takes a few minutes to cool down.

All this changes when you're talking about a road or rally racing car. The throttle percentages above are reversed—90 percent of the time the driver has his or her foot to the floor, and only 10 percent of the time

is the turbo not producing boost. That means that the heat being pumped into the intercooler cannot be removed fast enough. It quickly heats up and stays at a temperature determined by the cooling airflow and temperature of the boost air. This is where a highly efficient intercooler really shines, and it explains why race cars have large, expensive intercoolers mounted in the best possible location for cooling airflow.

A second measure of intercooler efficiency is the boost pressure loss between the inlet and outlet. The less pressure is lost, the better, since that means the turbo doesn't have to work as hard to make a given level of boost in the intake manifold. This makes the turbo more efficient by dropping the pressure ratio, and it means more power to the wheels instead of lost to heat.

Most street-use intercoolers are going to act as a heat sink rather than a radiator, so the best reason for upgrading the intercooler is to reduce the pressure drop. A larger intercooler also has more thermal mass, so it takes longer to get hot, absorbing more charge heat in the process. It does take longer to cool off, which is the downside to greater thermal mass. The tendency to get hot and stay hot is known as heat soak.

Intercoolers can be placed just about anywhere on the car, but to minimize plumbing length (which adds to turbo lag) most are mounted in the engine compartment. All Mitsubishi intercoolers are mounted either in front of the radiator or behind the headlight. The more direct airflow the intercooler gets the better, which is why high-powered race cars have their intercooler up front, like an Evo.

Intercooler Construction

The design of the intercooler core is very important for airflow and heat transfer ability. There are a lot of ways to create an intercooler core with sufficient area for airflow, as well as enough surface area for good heat dispersion, but only two have become commercially popular.

The first, called a tube-and-fin core, is made the same as an engine-cooling radiator. These cores use thin-walled, oblong aluminum tubes brazed to a header plate on both

In a bar-and-plate intercooler, aluminum fins are brazed inside and between the square aluminum "boxes" or "channels," similar to tube-and-fin intercoolers. The square aluminum channels are larger than the equivalent tubes in a tube-and-fin intercooler, and can fit more fins inside.

Intercooler end tanks are almost as important as the core in determining airflow, but packaging constraints usually determine which end tanks can be used. The ones shown here are used by RRE to fabricate different custom tanks, depending on the car they're to be installed in.

ends of the intercooler. Some tube-and-fin intercoolers (like the ones installed in Mitsubishi Starions) make use of extruded aluminum tubes with integral fins.

The second type of intercooler is bar-and-plate type. In this kind of intercooler, the lengthwise tubes are actually fabricated from strips of aluminum running down the intercooler.

Comparing the two types of intercooler cores, the external aerodynamics of the tube and fin are better in many cases, but not all. What this means is that cooling air from outside is better able to flow through the intercooler core and keep the intercooler temperatures down. This can make them more efficient in steady-state conditions, but there are many other factors that have a greater influence on efficiency.

In addition, the thin tubes and the method of brazing to the end tanks means that the internal aerodynamics of this kind of intercooler are not very good, which increases their boost pressure drop. Also, the many fins and overall light design of tube and fin intercoolers make them more susceptible to damage.

Bar-and-plate intercoolers are more robust than tube-and-fin intercoolers so they are more resistant to damage. They are not as aerodynamic externally, but they have a larger number of internal fins and larger internal flow passages than tube-and-fin intercoolers. This means that the pressure drop across a bar-and-plate core will be lower than the pressure drop across an equivalent tube-and-fin intercooler. Despite their higher cost, most aftermarket intercoolers are bar and plate because of these advantages.

Intercooler Spray

An intercooler spray system is a way to reduce intercooler temperatures. Sprayers are used on many recent turbo cars, including 2003–2004 USA-market Evos and most Evos sold in other markets.

A quick spray of water has the potential to drop intercooler temperatures significantly, depending on the temperature and humidity of the outside air and the charge air coming into the cooler. An effective intercooler spray system is controlled electronically. Like a good boost control system, it should be able to "anticipate" when the next burst of hot compressed air is going to raise the intercooler temperature.

Simply tying intercooler spray to boost won't work very well on a street car. By the time the intercooler temperature gets too high, the burst of boost is over, and so is the water spray. Intercooler temperature is a better variable for controlling spray, since peak intercooler temperature will not always coincide with peak boost.

One of the best cheap intercooler spray systems (if your ECU isn't equipped with intercooler spray control) is also the simplest: "driver control." Add a simple, cheap temperature gauge to your intercooler core, and a momentary switch to the spray system. When the temperature starts to increase, the driver manually toggles the intercooler spray. It's not perfect, but it may work better than a boost-controlled spray system.

An intercooler sprayer system is actually very easy to build. Use mister nozzles from the hardware store to blanket the intercooler core, and a cheap aftermarket windshield washer pump and reservoir to drive the system. Add a relay and a temperature gauge and you'll get very good results for a low price. If you want to step it up, look for one of the aftermarket sprayer controllers to add some finesse to your setup. Be sure however that you don't accidentally reduce intercooler performance by restricting the cooling airflow with sprayers and hoses.

Every once in a while the aftermarket comes up with a variant of the intercooler cooling spray system using nitrous oxide or other compressed gas (usually CO_2) to cool the intercooler. Such a system can actually cause the intercooler to freeze over with ice from water in the air, but it's an expensive way to get minimal extra benefit. Sure, the greater temperature drop is good for intercooler efficiency, but the added weight, cost, and complexity of a compressed gas system make it hard to recommend. It may also bump you up a class or two in competition if it's counted as nitrous injection. Use compressed gas to cool the intercooler before a drag run if that's your thing, but don't feel tempted to spend that much cash on an ultra-cold setup.

The same goes for using ice in your intercooler spray bottle. As it turns out, most of the benefit of intercooler spray is from the water changing state from liquid to gas. A few degrees difference in water temperature (for example from adding ice) does almost nothing to decrease the intercooler surface or outlet temperature. You'll get more benefit from using a finer nozzle and higher-pressure pump to reduce droplet size as much as possible. The smaller the droplets, the quicker they evaporate from the surface of the intercooler, and the quicker the temperature drop is realized (though the absolute temperature reduction is no different).

Chemical Intercooling

Water/Alcohol Injection

Water injection seems to be one of those power-adding technologies that come in and out of popularity every few years. But it's not new; the basic operation and effects of water injection was well understood in the 1940s, during World War II. Henry Ricardo, one of the most influential piston engine designers, is credited with first applying water injection to piston engine fighter aircraft to help them squeeze more power out of the engines on takeoff.

There are several reasons that water injection works. First, as the fluid enters the hot charge air stream, it absorbs heat from the air and this cools the charge before it enters the combustion chamber. Although the water displaces some of the air that could be used to make power, it more than makes up for it (up to a point) because it increases the density of the air going into the combustion chamber. Next, inside the chamber, the water vaporizes with the heat of combustion and further reduces the tendency of the engine to knock by reducing local temperatures in the combustion chamber and helping keep overall temps down.

Ricardo and other experimenters later found that alcohol, particularly methanol, was even more effective than pure water at suppressing detonation. Since the alcohol burns in the combustion chamber it adds to the power production of the engine. Of the available alcohol choices, methanol is preferred since it has a high octane rating and adds to the thermal effects of injection. Alcohol is also less likely to freeze in cold weather, which is a factor for a street-driven car.

Pure alcohol is flammable, poisonous, and difficult to handle and transport, but it's been found that 50 percent water and 50 percent alcohol mixtures are nearly as effective as pure alcohol. It helps that you can

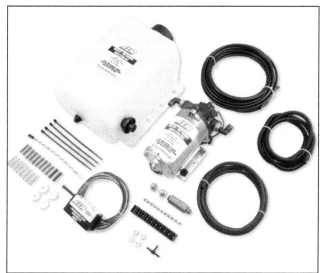

Properly used, a good water injection kit like this AEM unit can overcome the disadvantages of running street fuel on a pumped-up turbo engine. Even western states' 91-octane fuel can be made to produce excellent power numbers. It's an easy 10 to 15 percent power increase over no water/ methanol injection.

easily find 50 percent alcohol/water mixtures in your local department store in the automotive section (most summertime windshield washer fluids are just that with a little dye added).

Interestingly, some of the effects of chemical intercooling can be gained by simply richening up the mixture and dumping in excess fuel. While the engine may only need a 13.0:1 or 14.0:1 of air-to-fuel ratio for optimal combustion, stock ECUs often dump excess fuel up to a ratio of 8.5 or 9.0:1 air/fuel at high loads. The extra fuel doesn't burn, but it does help cool the combustion chamber, reducing detonation tendencies and thermal stresses on the pistons, valves, exhaust, and catalytic converter. Unfortunately, this is an inefficient way to cool the combustion chamber.

So if it's such a good source of "free" power, why don't OEM manufacturers use chemical intercooling? The downside of water injection is that it adds a maintenance hassle in refilling the system, and requires some sort of failsafe that will disable boost when it's not able to inject. Both of these reasons, and the cost of a good, effective system, are why OEM turbo cars don't come equipped with a water injection system.

Tuning Water/Alcohol Injection

If you decide to experiment with water/methanol injection, buy a reputable kit. You can build one yourself, but the kits come with better quality pumps, lines, and controllers than most home-built systems. Here's a hint: You can't use windshield washer pumps, which are fine for intercooler sprayers. A pump with enough flow and pressure for a decent water injection system is not cheap, but it is worth it.

The best kits include boost switches and some method of reduc-

ing or turning off boost if the system fails. The simplest failsafe is a fluid level switch connected to a solenoid that bypasses the boost control system. When the level falls below a preset point, the system allows only wastegate boost. But consider also what happens if a nozzle is clogged or the pump fails, as a simple level-sensing system will not react to such a situation. Some injection system manufacturers have a flow-sensing failsafe, but it's up to you to decide if the extra cost is worth the extra peace of mind.

The mixture of air and water/alcohol can be progressively varied with boost or airflow, like fuel, but it's not really needed. Unlike the air/fuel mixture, the water/alcohol/air mixture has a wi de range of useable ratios. The supplier of your water injection kit will be able to recommend a specific nozzle size for your engine, but as a starting point, most tests show the most effective range to be 10–15 percent of the fuel being injected. So if you're running 550-cc/min injectors at an 80 percent duty cycle on boost, your water injection should be flowing around 200 cc/min (.10 X ((550 X 0.8) X 4)).

Tuning for water injection is not hard. Most tuners set up the water injection system so that it comes on 100 percent as soon as the engine reaches full boost. If the system is staged or progressive, it's usually set up to come on with the onset of boost (around 5 psi) and increase to 100 percent by the time full boost occurs. If you implement a failsafe (a good idea), make sure that your timing map is safe on wastegate boost (usually 11–15 psi).

That shouldn't be a problem with most 4G36t timing maps, as even the stock timing maps are conservative enough for wastegate boost. Add timing only in the high-boost areas of

the map, when the engine truly needs more octane. When the system is working, the higher boost levels will enable more power production without a change in ignition timing. When it's not working, the wastegate boost level and a light on the dash should be enough to save the engine from an inattentive driver.

Bypass Valves and Blow-Off Valves

All that plumbing between the turbo, intercooler, and intake manifold adds up to a pretty large volume that has to be filled with pressurized air. The time it takes to fill that space adds to lag, but once it's filled, all that air has a lot of inertia. It doesn't like to be started moving or stopped suddenly, like when you bang the throttle closed during a shift. If there was no way to prevent it, each shift would cause the entire mass of air to stall momentarily until the throttle is opened again.

All that air that slammed into the back of the closed throttle plate would send a shock wave back to the compressor wheel. The sudden pressure difference between the compressor and turbine wheels causes an imbalance that can destroy turbo bearings. The sudden change in compressor flow also throws the compressor into the inefficient surge area, which heats up the charge air.

Bypass valves (BPVs, sometimes called diverter valves or recirculating valves) and their cousins, blow-off valves (BOVs or dump valves), solve this problem neatly. BPVs are installed near the throttle plate, and have a diaphragm actuated by pressure inside the manifold. As soon as the pressure in the manifold drops (as it does when the throttle is slammed shut), the valve opens and allows the pressure and shock waves

The Evo VIII and IX BPV is a solid part that's good to 25 or more PSI of boost. You shouldn't have to replace it until you're really testing the limits of the stock short block, and well past the limits of the stock turbo. The similar-looking 2g BPV, on the other hand, is a guaranteed boost leak.

in front of the throttle to return to the turbo inlet.

Because a BPV returns the air after the MAS, it has already been measured by the ECU and there is no change in air/fuel ratio. As hinted at in the bolt-ons chapter, the 1g and Evo BPVs are very good and don't need to be replaced for moderate-boost applications (under 20 psi) unless you want something that looks or sounds a little nicer.

The 2g BPV, on the other hand, should be ditched as soon as possible. Replace it with a used 1g or Evo BPV at the very least, or a decent aftermarket BPV if you want. At really high boost levels, any of the stock BPVs will start to flutter and cause odd boost variations that can be hard to tune around. A tight aftermarket unit will get you around this particular problem, but don't rush to slap one on unless you have a good reason to suspect boost is leaking around your current BPV.

BOVs, which vent the excess pressure externally, cannot be used on a stock MAS setup without undesired drivability problems. Since the air that is blown out during a shift

MHI Turbo Spotter's Guide

Mitsubishi Heavy Industries is a part of the gigantic Mitsubishi company, which includes Mitsubishi Motor Company as well as Mitsubishi electronics, shipping, banking, and dozens of other industries. MHI turbos are classified by family, size, and compressor design. The number/letter "alphabet soup" used to describe them can be confusing.

The most common turbos for automotive use are the TD04, TE04, TD05, and TD06 families. Each is larger (in housing size and shaft diameter) than the one below it. The center sections and turbine housings of each family are not interchangeable, although some aftermarket turbo companies have engineered or modified parts to create hybrids of TD05 and TD06 parts, for example.

When a turbo such as the TD05H-14B-6 is called out, the family suffix "H" indicates one style of turbine housing. If you see an "R" in the name, it's the code for reverse-rotation turbos, like the TD05HR turbos used on the Evo IV through IX. An "A" after the family designation indicates the aluminum/titanium center section installed on Japan-market Evo RS models, such as the TD05HRA-16G6-10.5T.

The number/letter combination after the family designation refers to the compressor wheel. The number is the diameter of the wheel, while the letter refers to the blade style and number. The Evo III and later 16G6 compressor is different from the regular 16G wheel in the size of the inducer (small diameter) of the compressor wheel.

The final number suffix designates the turbine inlet area, such as the TD05HR-16G6-9.8T installed on the USA-market Evo VIII and the TD05HR-16G6-10.5T installed on the Evo IX, respectively. The "9.8" and "10.5" refers to the area of the turbine inlet in square centimeters. Finally, the "T" suffix after the turbine inlet size designation refers to the turbo having a twin-scroll exhaust housing.

Unfortunately MHI doesn't see fit to publish A/R ratios for their turbos, so there's no way to directly compare a particular turbine inlet with another manufacturer's turbo. This is true in general, too—there's no good way to compare turbines without published A/R ratios.

Some common MHI turbos and their rated maximum airflow capacity at 2:1 pressure ratio. We've added the Garrett T-25 (stock 2g) and T-28 (aftermarket bolt-on 2g) for comparison:

Turbo	Model Number	Approximate Compressor Airflow @ 2:1 Pressure Ratio
Stock 1g automatic	TD04-13G-5	360 cfm
Stock 1g manual	TDO5H-14B-6	430 cfm
Stock 2g Manual/Auto:	Garrett T25	330 cfm
Aftermarket 2g bolt-on:	Garrett T28	400 cfm
Stock Evo I-II	TDO5H-16G-7	505 cfm
Stock Evo III	TDO5H-16G6-7	550 cfm
Stock Evo IV-IX	same as above, but reversed blades	
Aftermarket 18G	TD05H-18G-7	590 cfm
Aftermarket 20G	TD05H-20G-7	650 cfm

has been metered, the engine will spike rich for a second as the ECU sprays extra fuel to match the air. Since the air is no longer there (it's vented outside the system), the excess fuel is sent out the exhaust unburned.

The BOV is the BPV's racier cousin. Everyone likes the sound of a BOV, since they dump the charge air directly to atmosphere with a big whoosh, just like on a real race car. Unfortunately they can't be used easily with the stock ECU.

During a shift this isn't a big deal, but as you come to a stop the opposite problem will occur—the BOV will pop open, and the car will stall from running lean as it sucks in unmetered air through the BOV. BOVs can be used, but they require either a conversion to a MAP-based (Speed-Density) engine management system, since there is no air to meter, or a conversion to blow-through MAS mounted after the BOV, either of which is a complicated fix just to achieve the BOV sound. When used in conjunction with metal intake and upper intercooler piping, a high-flowing aftermarket BPV (in this case a BOV set up to recirculate the air) will give you a louder sound than stock, without the drivability problems.

Stock Turbos and Intercoolers

The DSMs and Galant VR-4

There were at least four turbos installed on the USA-market 4G36t between 1989 and 1999. The 1g manual transmission cars and Galant VR-4 got the best turbo; a MHI TD05h 14B with a 6-cm^2 exhaust housing. It's a good match for the engine and has a lot of potential for increased power.

The 14B will happily hit 15 psi of boost on a street DSM, or 18 psi with an upgraded fuel system, although it becomes a limiting factor at around 17 psi of boost or about 250 hp at the wheels. This is a good street turbo for both the 1g and 2g DSM cars if you can find a good used one, although they are getting old and harder to find than a good aftermarket bolt-on turbo. 1g automatic cars got a smaller MHI turbo, a TD04 hot side with a 13G compressor. This turbo is barely adequate for stock horsepower levels but it does spool at a very low RPM. This helps the automatic get off the line with a big boost of torque down low.

You're looking at the smallest of the MHI turbos delivered on a USA-market 4G63. It's the pathetically small 13B from a 1g automatic car. Luckily it can be replaced by a "regular" 1g turbo (from a manual transmission car) with a new exhaust manifold and some small plumbing changes. Don't waste your time trying to squeeze power from this turbo.

The 2g DSMs got shafted on this one. Instead of the good-sized MHI turbo installed on the 1g, the 2g manual got a Garrett T25 turbo that is actually just barely adequate for the stock horsepower rating. It spools very quickly, but it will not hold much boost at high RPM. Most people replace it as a matter of course. Note that the T25 does not share the same outlet position, oil, or water lines as the MHI turbos. This necessitates an install kit when you're swapping out your T25 for a 14B or other MHI turbo.

Intercoolers for all of the DSM triplets (Talon, Eclipse, and Laser) were smallish and mounted between the front tire and headlight. So-called side-mount intercoolers like these are notoriously inefficient, because there just isn't much cooling airflow in that area. Only when you get up to freeway speed does your stock side mount start to do any useful cooling. The plumbing is also rather restrictive since it has to move the air from the front-mounted turbo to the side of the engine bay while avoiding any accessories hanging off the front of the engine.

In theory, the Galant VR-4's front-mount intercooler should be the best of the bunch. It mounts up front before the radiator just like an Evo's intercooler, but the similarity ends there. The GVR-4 intercooler is pathetically small and restrictive. It also has some awfully restrictive plumbing to snake the air down under the radiator and up front to the cooler. While it has a good location, the pressure drop and heat soak tendencies of this intercooler make it worthless for performance use. Almost anything is better than the stock GVR-4 intercooler.

Stock Evo Equipment

As with everything else in the 4G36t world, the USA-market Evo

VIII and IX got the best turbo, intercooler, and plumbing of any production car. That doesn't mean that there is no room for improvement, however. The turbo, while bigger than any DSM turbo, is sized to emphasize engine life and warranty durability over high power.

All models from the very first Evo have been equipped with a variant of the TD05h turbine and housing with a 16G compressor wheel. From the Evo III on, the turbo incorporated a larger wastegate and larger compressor wheel. With the advent of the Evo IV and on, the turbo rotation was reversed to make plumbing easier in the engine bay (since the transmission was "flipped" compared to Evo I through III models).

The USA-market Evo VIII came with a twin-scroll variant that uses a special split exhaust manifold to direct the pulses from number-1 and number-3 cylinders to one half of the scroll, and the pulses from number-2 and number-4 to the other half. This makes it an incredibly responsive turbo good for full boost below 3,000 rpm on a stock 4G36t. The large 16G compressor means it's as good as any other big 16G turbo for high-airflow boost, and it will easily produce 300 or more horsepower to the wheels.

The Evo VIII has a unique, reverse-spinning variant of the 16G turbo. Its MHI code is 49178-01560. Notice how the compressor blades are folded over to the left, opposite the blades on the Evo III 16G shown above.

The ultimate 4G36t, the Evo IX, came with a larger variant of the same turbo designed to keep up with that engine's better-breathing MIVEC cylinder head. The biggest difference between the Evo VIII and IX turbos is the hot side. The turbine housing on the Evo IX turbo has a 10.5-cm² inlet, resulting in a larger, more free-flowing A/R ratio.

On the Evo IX engine, this turbo also spools earlier than the Evo VIII thanks to the engine's MIVEC increasing low-rpm airflow through the head. It's a great VIII upgrade turbo, with the downside of losing a few hundred RPM in spool in cars without MIVEC.

The Evo has a generous-sized intercooler, but the stock intercooler core is compromised to fit easily within the bumper. In addition, it is somewhat restrictive at high power levels.

The size and exact dimensions of the Evo intercooler have changed over the years, but it's always been mounted in front of the radiator, below the front bumper. It's a tube-and-fin intercooler and gives a fairly large pressure drop. At any point above 350 hp at the wheels, the Evo intercooler is a restriction; it can be a power limiter for most large turbos.

Stock Turbo Modifications

Once you have the bolt-ons taken care of, and your boost level is set to a moderate, but high level, you might be ready to start changing your turbo around. But before you go swapping out turbos, there are a lot of things that can be done to wring quite a bit more power out of your stock turbo (yeah, even the 2g folks!).

On the exhaust side, stock DSM turbos have a sealing ring that isn't really necessary, and in fact restricts flow out of the manifold and into the turbine housing. Start by cutting away the step and matching the turbo to the manifold opening.

You can maximize your existing turbo by porting the intake and exhaust sides. You won't be able to go too far into the scroll, but with a steady hand anyone can make a noticeable improvement. You can also send out the stock turbo to a shop for professional porting work. Some turbo shops build hybrid turbos using parts from different turbo models. Depending on the specs, a hybrid turbo might be a way to get stock spool with more flow on top. The 2g replacement T28 is one turbo that is well-liked by autocrossers and street drivers who don't want to give up the instant spool of a small turbo. This also means you do not need an install kit, since the T28 is related to the T25 and uses all the same mounting points and feed lines.

Another service performed by turbo shops is "clipping" the exhaust

Porting the stock turbo exhaust inlet will definitely help with high-RPM airflow, but don't expect huge gains. In fact, it may even hurt spool since exhaust gas velocity won't be as high as before. Start by porting out to the edge of the step present in the stock turbo inlet, and match the manifold to the same diameter. Leave a fine polished finish so carbon can't build up as fast.

wheel. This involved turning down the outside diameter of the turbine wheel so that it displaces less of the turbine housing. This raises the spool RPM by making the turbine a little less efficient, but the benefit is that the exhaust turbine can now flow more freely. This allows the engine efficiency (and boost) to stay up at high RPM and airflow levels. It also helps to bring down the RPM of the turbo itself, keeping it within its efficiency range. The ideal would be to swap the exhaust turbine for one with a larger A/R ratio, but clipping is a discount substitute.

Turbo Swaps

If you've done all the bolt-ons, pretty soon you'll be running it at an inefficient level of boost and airflow that will cause the compressed air to become even more heated than usual, or your engine will be making more power (and consuming more

air) than the turbo can supply. In this case, boost will fall off at high RPM. The point of inefficiency varies depending on which 4G36t we're talking about, but is between 220 hp (2g DSM) and 350 hp (Evo IX). Once you reach this power level, there isn't much you can do other than swap the turbo. Dollar for dollar, you'll get more horsepower out of a larger turbo than any other modification you can do.

But believe it or not, there are reasons to not swap your turbo for a bigger one. With the possible exception of the 2g's T25, your stock turbo is sized about right for getting the most useable power on a daily basis. Going up a size or two will lose hundreds of RPM of spool. If your car is 100-percent street driven, or if you like to compete in autocross races, a bigger turbo might have downsides that make it less fun than a stock turbo. It will definitely make your car slower in some circumstances; for example, if your gearing is too high to make use of the raised power band from a large turbo.

In addition, all larger-than-stock turbos require some kind of tuning. Some (like a 2g T28 swap) can be done with nothing more complicated than a boost controller and a careful eye on the EGT gauge, but most larger turbos will at least require that the fuel system is upgraded. The stock 1g fuel system, for example, cannot handle even a slightly larger-than-stock turbo. It's pretty simple: If you don't want to upgrade your fuel system, intercooler, and exhaust at the same time, don't go swapping turbos.

Other Stock Turbos

At some point, the stock turbo will become your bottleneck and you'll be ready to step up in turbo size. If you've set your sights on a bigger turbo and more power, there are

a few options. The first step to getting more airflow is to make sure that you're running the best stock turbo that you can. An easy power swap for your automatic DSM is to upgrade to a standard 1g manual turbo. The manifold and oil lines have to be changed at the same time. The automatic TD04 has a smaller bolt pattern, and the oil lines are slightly longer for the larger turbo.

The same goes for the 2g—swap to a 1g manual turbo for a budget power increase. The 14B is a good all-around turbo that will allow you to make decent power without spending much money. It's not hard to swap over, and most of the parts that you need to do a 14B swap into a 2g, for example, are the same parts you will need if you later decide to go bigger.

If you're running an Evo VIII, get an Evo IX turbo for the increase in high-end power (and lag). It's almost a bolt-on; just a few parts are needed to swap. The resulting power increase is worth it if you get the turbo cheaply, but probably not worth it if you have to pay full list price for a new one from Mitsubishi. Have your car tuned and enjoy 350 peak horsepower and streetable lag. The IX doesn't have a readily upgradeable replacement turbo, except maybe the ultra-light-weight titanium/aluminum unit from the Japanese-market IX MR. It won't flow any better than the stock turbo but it will spin up faster thanks to its low rotating inertia.

The Evo VIII and IX turbos (as well as turbos from the Evo IV, V, VI, and VII) are not bolt-on for a DSM. They are reverse rotation. This means they are designed like a mirror image of a DSM turbo, like everything else on the Evo, so they will end up upside down on a DSM manifold. You could build a custom manifold and plumbing to use one, but it's probably not worth the effort.

The TD05h-16G turbo family is a good bolt-on turbo for the DSM cars. If your stock DSM turbo is dead, you'll be happy to find out that an aftermarket 16G turbo conversion complete with oil lines, turbo, and all necessary parts is several hundred dollars cheaper than a stock 14B turbo center housing from Mitsubishi, and about the same price as an aftermarket stock replacement turbo.

Any of the 16G turbos makes an excellent street turbo, which makes sense since they were all originally used on stock 4G36t applications in other markets. The largest of the family, the famous Evo III 16G shown here on the left, has an MHI code 49178-01470 on the compressor. It's a 350+ hp turbo. The regular "big 16G" on the right has an MHI code 49178-01420. The difference between the two is in the exhaust housing (the inlet is larger on the Evo III exhaust housing). It will flow more exhaust gas at high boost levels, preventing exhaust manifold pressure from rising as high.

Hardcore DIY-ers might consider buying a 16G wheel and compressor housing with a rebuild kit for your existing 14G center section. The problem is balance. Turbo shafts and wheels are very precisely balanced (that's what the numbers on a new wheel are for) and a garage-built turbo has very little chance of surviving on random balance at hundreds of thousands of RPM.

If you really want to save money on your 1g build, you can even buy a 16G center section with the compressor housing attached and keep your stock turbine housing and actuator. This only applies to the smallest 16G turbo, and your high-RPM power will be limited by the low limits of the stock exhaust housing.

The biggest of the 16G turbos, the Evo III 16G with a 7.5-cm hot side, is the star of the lineup. Since they all cost about the same, the only reason to pass on the Evo III variant is the slightly greater lag that it will give you, or if you want to swap compressors while keeping your stock turbine wheel and housing.

If you can stand having full boost come on as high as 3,300–3,500 rpm, it's a good choice. Ultimately it will be good for up to 320 to 350 hp at the crank. That's maybe 300 at the wheels (or as much as you would want with a stock DSM drivetrain).

You won't win many "dyno drags" with an Evo III 16G on a DSM, but if it's well tuned it will make your DSM a wonderful street/track terror that's capable of getting deep into the

12s in the 1/4 mile. These turbos are very tough, another good reason to buy one. Besides the increased lag from stock, the other downside to the Evo III 16G is its tendency for boost creep. This is mostly a problem with boost levels under 20 psi using a 3-inch cat-less exhaust. If you port the turbo, keep a cat in the system, and tune for 20-psi peak boost, you will be very happy with your choice.

Bigger Bolt-Ons

If you're building a street/strip or autocross car, chances are one of the better stock turbos will be all you need. You can go bigger, but the lag from a large turbo is something that you may not fully appreciate if you plan to drive your car every day.

Autocrossers will be very disappointed in anything bigger than a 16G, because it's unusual for a course to have a straight long enough to reach full boost for very long with a big turbo. On the street, having a huge hole in your power curve from 2,000–3,500 rpm is not fun, and you might be happier with a smaller turbo that spools better.

This big turbo for a DSM engine is bolt-on, but just barely. Notice that it requires an external wastegate, and a custom matching O_2 housing (which actually doesn't include an O_2 sensor). This replaces the stock turbo but requires some fabrication to make sure all the plumbing lines up correctly.

That said, if 300 semi-reliable horsepower won't satisfy you, and if you're willing to give up the lower part of your rev range, you can go for the big power numbers with a bigger turbo. Getting 400 hp out of only 2.0 liters requires a lot of airflow, though, and a lot of revs. This takes its toll on engine parts. Unless you have an Evo long block in nearly perfect condition, you should proceed from this point with extreme caution. Once you've reached the 300 wheel-horsepower mark, things get expensive (and unreliable) quickly.

The margin of error for tuning a 350-hp DSM motor is much narrower than a 300-hp motor. Your pistons might be just fine (in fact there are many, many 400-hp DSMs running around with stock pistons in them), but they might not be. At this power level, a little knock can destroy lots of internal engine parts. High boost can blow head gaskets, and this much power will likely damage the stock DSM axles, driveshaft, and differentials. You've been warned.

If you have an Evo, the 350-hp range is not nearly as on the edge. In fact, you should be able to make nearly 400 hp in short bursts before you run into the kind of weakness that a DSM motor exhibits at 350 hp. If you plan to make above 300–325 hp regularly (if you do any track driving) with either a DSM or an Evo, you should spend the money to build a tough short-block.

Despite all the fancy aftermarket hybrid turbos created from parts of this-n-that, there are a couple of old standbys that have rocketed many DSM and Evo drivers deep into the 10-second range in the 1/4 mile, with as much as 400 hp. That would be the MHI 18G and 20G turbos, with various hot sides depending on the application. Both of these turbos are pretty dated for real high-power

These days, there are lots of great choices for big-power turbos that bolt to the stock, or stock replacement, exhaust manifold. Any big turbo that uses the stock manifold and downpipe is a "bolt-on" upgrade, although some turbos in this category require some pretty serious modifications to the cold side plumbing as well as the water and oil lines.

use, but they are still widely available and common.

Of the two "big" MHI turbos, the 20G is arguably the better one. The 18G's 590-ctm flow rating is only about 10 cfm more than the Evo III 16G turbo's flow rating, which isn't enough of an advantage to make up for the 18G's vastly increased lag. By using an 18G instead of an Evo III 16G or similar you'll be giving up 500 rpm of streetable spool in exchange for maybe 10–20 hp, depending on several factors. On the other hand, install a 20G on your DSM or Evo and you'll be looking at a potential 400 hp at the wheels with about the same spool as the 18G. There's an old DSM saying: "The 18G has all the lag of a 20G with all the power of a 16G."

Most 20Gs for both DSM and Evos use a single-scroll turbine housing to reduce exhaust restriction, although this adds lag to the setup. Hybrid Evo 20G turbos are also available, which combine the stock twin-scroll Evo VIII or IX exhaust housing

and center cartridge with a 20G compressor wheel and housing. It's pretty hard to argue for the 20G as a streetable turbo, but many thousands of them are driven daily. If you plan to put street miles on your Evo, go for a 20G hybrid because you can use all the quick spool you can get.

Non MHI Bolt-on Turbos

Modern turbo compressor designs from Garrett, Holset, Schwitzer, and others have largely taken over from the 20G in many applications. The aftermarket is filled with companies creating their own big bolt-on turbos using custom MHI-compatible turbine housings with various combinations of turbine wheels and compressors.

Aftermarket companies like APS, Forced Performance, and Greddy create turbos designed for big power numbers while being easy to install and tune. The best ones of the group give more power on top than the dated MHI compressors without sacrificing any spool over a stock or 16G-style turbo, and are as easy to install as a stock replacement. Forced Performance's FP Red and Green turbos are examples of bolt-on, non-MHI turbos, as are the various Blouch turbos and the PTE Schwitzer.

Most of these aftermarket turbos are really good deals, and can produce all the power you want. Want a ball bearing for fast spool? Done. Need 500-hp capacity? No problem. Just don't get too attached to the concept of bolt-on, since it can mean different things for different people. If you want to go this route, look for a turbo that requires fewer modifications to run—the more complete the parts kit that comes with it the better. You don't want to be in a position of having to track down odd adapter fittings to mate with non-standard threaded parts.

Some of the so-called bargains in bolt-on turbos actually come almost bare. If you buy one of these, be prepared to figure out your own water and oil feed and return lines, intake adapters, and possibly even new intercooler plumbing. Stick with one of the reputable suppliers like those mentioned above if you want to keep the install straightforward.

One thing that makes shopping for a big bolt-on turbo hard is the lack of information published by some turbo re-sellers. Since these turbos are all created from the same parts catalogs of housings, compressors, and turbine wheels, the builders/re-sellers are reluctant to release specific information like what parts are used and what flow maps apply to their turbos.

It's a competitive business and these companies are doing their best to prevent competitors from copying their particular combination of parts and machine work. Sometimes they publish compressor maps and airflow/horsepower ratings, sometimes not.

Unfortunately this leaves us as consumers in an information vacuum. About the best you can do is talk to the manufacturer and get as much information as you can, and then compare that to what you read on the Internet. Use dyno printouts and drag race time slips as unbiased sources, and don't forget that most people won't tell you that the $2,000 turbo kit they just installed did not pay off when it came time to hit the track.

Beyond the Bolt-On

Bolt-on turbos and turbo kits can take you as far as most of us want to go, or rather can afford to go. For really stratospheric power levels, and for true track-only time attack or drag race cars looking for 500 hp or more, this might not be enough. About 99 percent of street car builders will stop

at this point because things can get very expensive very quickly. However, if you really must know all of the performance characteristics of your next turbo, you probably won't be able to find a bolt-on turbo that is fully documented.

Off-the-shelf big turbos don't usually come with the bolt-on Mitsubishi flange, which means you'll need a new exhaust manifold. If you're talking about a race car, making or buying a custom exhaust manifold is a small part of the overall project cost, but it's hard to justify for a street or street/strip car.

Swapping turbo styles will require some other extensive mods, so you should be prepared for them. The output flanges are not compatible with the stock O_2 housing and downpipe either. There just isn't

room within the confines of the Mitsubishi outline for the flow that a really big, 500+ hp engine requires.

Probably the most popular flanges for really big turbos are the Garrett T3 and T4 flanges used by many domestic manufacturers. Going to a T4 flange blows the doors open to a huge variety of turbos from many different manufacturers, including Holset and, of course, Garrett. These turbos should satisfy the power urge to unheard-of levels.

Whatever you do, make sure you read this chapter to understand how to choose and evaluate a bigger turbo, and how to know when you need one. Read some other books, too, and talk to the winners in your racing class. Copy what they do for starters, and change your setup only when you're able to get close to them.

Building or buying a custom turbo manifold means you don't have to stay within the limits of turbos that replace the stock one, so you might also consider this route if you want to experiment with different turbos to tune the characteristics of your engine's boost response. Be prepared for the amount of fabrication involved, though.

Turbo Swap Collateral Damage

Don't forget, if you're working on a DSM, that these cars are quickly approaching 20 years of age. That means a couple of things for the enthusiast. First is that you can't assume your car has the stock turbo on it. Even if you think your car is stock, make sure it is before ordering parts. Get a good look at it and make sure you're looking at what you think you're looking at. Over the years many 2g DSMs have had their stock 1g turbo replaced with a 14B. Many 1g 14B turbos have been replaced by 16Gs and other smallish bolt-on turbos. Many, many downpipes and O_2 housings have been swapped over the years too, so make sure that they're stock or proven aftermarket before you buy a new turbo.

Second, be prepared for a fight if you decide to swap turbos. The heat and exposure to the elements corrodes exhaust fasteners badly, and you may not be able to get them all out without breaking a few off. Re-read the broken bolt section of the bolt-ons chapter (Chapter 2) if you need a reminder. The bolts between the manifold and turbo and manifold and head are notorious for breaking off. Spray them liberally with a penetrating lubricant when cold (not WD-40; Kroil and PB Blaster are some of the better penetrants we've used). You might even do this a few times in the week before you plan to swap parts. Make sure you have an alternative to getting your car running the same day if it's a daily driver, and make friends with a local machine shop in case you need some bolt removal services.

This is also the perfect time to replace your exhaust manifold.

Check it thoroughly for cracks (if you have a 1g with the stock manifold you can guarantee it's cracked). Always replace the manifold-head and manifold-turbo gaskets with new OEM parts when you disturb their seal. Replace the turbo-manifold bolts every time you remove the turbo to prevent broken bolts the next time you do the job, and use anti-seize on every bolt on the exhaust system.

All of the coolant, oil, and air lines connected to the turbo will likely have gotten hard and may crack or tear when you remove them. Always replace rubber hoses when you're in there—the coolant line across the front of the block, behind the exhaust manifold, cannot be accessed any other way. In most cases, replacing all these lines and hoses will cost more than the turbo does, but it doesn't make sense to put 18-year-old hard rubber hoses back on when it's so hard to replace them.

Always replace the turbo oil feed line when you install a new turbo. If you're tight on cash, flush it out

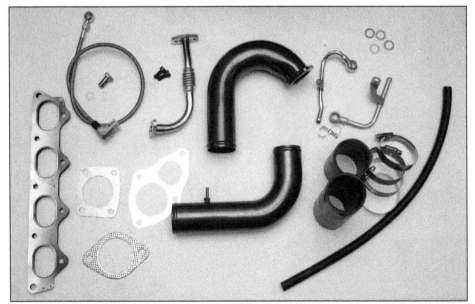

There are a lot of small parts that have to be removed and replaced or modified when doing any turbo swap, even a simple stock-for-stock swap. This is all the parts needed to swap a 2g T25 for a 14B or 16G turbo meant for a 1g, and it's a good introduction to the turbo system's extra parts. Included (clockwise from left) are exhaust manifold gasket, oil feed line and banjo bolts, oil drain line and bolts, upper intercooler "J" pipe, coolant hard lines and crush washers, coolant hose, silicone couplers and clamps, lower intercooler pipe, turbo manifold, turbo O_2 housing and downpipe gaskets.

When replacing the stock turbo, always replace the long bolts (sometimes two bolts and two studs) that hold the turbo to the manifold. Many times they break off when removed. Even if they don't break, they will be weakened from the heat. Use OEM Mitsubishi bolts if possible.

carefully with solvent. Oil feed lines often become filled with carbon deposits that will damage your new turbo, and even a cleaned used line is a liability. In fact, most aftermarket turbo suppliers will not warranty a turbo that was installed with a used feed line. It might be more expensive than a new stock line, but a braided stainless-steel -4AN Teflon hose will make future turbo changes easier and cleaner, as well as be cheaper to replace in the future once the adapters are in place.

Consider modifying your turbo oiling source, too. The problem with the 1g DSM turbocharger oil source on the head is that it is at the very end of the oiling system, about as far as you can get from the oil pump. That means that the pressure at this point will be very low. Also, any bits of dirt or wear particles present in the rest of the oiling system can migrate here. The result of both problems is unusually fast turbo bearing wear.

You can install an in-line oil filter to eliminate the contamination issue, but this does not solve the problem of pressure; in fact it makes it worse. The best thing to do is plug the original oil feed port on the head and take your oil from the oil filter adapter. Several aftermarket companies sell an adapter fitting that will connect the stock plugged port to -4AN hose. Finally, an in-line oil filter is not optional with a ball-bearing turbo; it's necessary. These bearings are more sensitive to dirt and grit, so always add a filter in the feed line to prevent early death.

If you're replacing your stock MHI turbo with another MHI turbo (like a 1g swap to a 16G turbo), you will be able to reuse your existing oil and water banjo bolts and possibly the hard lines that attach to them. For any other swap, prepare to modify your existing parts or buy new

Parts Needed When Installing an Evo IX Turbo on an Evo VII		
Quantity	Part #	Description
1	1310A171	Pipe, t/c Water Return
1	1310A248	Pipe, t/c Water Feed
1	1225A042	Tube, t/c Oil Feed
2	MF660063	Gasket, Clutch M/c (10mm crush washer)
2	MF660064	Gasket, a/t Case (12mm crush washer)
1	1515A057	Fitting, t/c Air Out (J-pipe)
1	MR281085	Gasket, t/c Air Out

ones. You may have already considered the manifold and downpipe; manifolds have to be swapped to go to one of the large stock-location turbos, for example, or from an automatic 1g to a manual 1g's 14B turbo. Downpipes are interchangeable, although O_2 sensor housings are not always the same. Make sure you have an O_2 housing (or exhaust elbow for an aftermarket turbo) that is compatible with the turbine outlet of your chosen turbo.

Ask the turbo supplier about other plumbing—make sure he tells you what size the oil and water feed fittings are. Many of the Garrett housings use 1/4- and 1/8-inch NPT for oil feed lines, and AN for the water feed lines, although this is not a rule. If you install a "dry" center section on your DSM, be sure to block off the water feed lines (1g) or add a bypass hose between the stock water feed and return lines. This is not a big deal, but something to watch out for.

Be prepared to deal with other plumbing fitment issues before you start on your project. Oil drain fittings are more interchangeable than they appear. Garrett drain fittings can be used on MHI center sections with a proper gasket and lots of sealant, and MHI drain fittings can be used on Garrett center sections by enlarging the bolt holes to 1/4 inch (from 6 mm). Keep your drain lines as straight as possible.

For compressor inlet and outlet plumbing you may have to use silicone couplers and mandrel-bent tubing sections to adapt the various parts of your setup together. The J-pipe that connects the turbo outlet to the lower intercooler pipe is the most common source of hassle. You can use a 1g pipe and turbo on a 2g by reversing and stretching the lower intercooler pipe, but it's not an ideal way to go about it. The easy way is to use an adapter kit with water and oil lines, as well as a J-pipe or modified intercooler plumbing. Some of your stock 2g lines can be reused, but a clean install requires a new oil line and oil drain flange at a minimum.

Installing an Evo IX turbo on a VIII is similarly simple, with the parts you need no further away than your local Mitsubishi dealer or salvage yard. The two cars are so similar you can use OEM parts. The manifold and intake plumbing can all stay the same, although you will have to replace the turbo air outlet J-pipe, as well as the oil and water pipes. The list above details the parts needed.

One last consideration with both bolt-on and non-bolt-on turbos is clearance between the turbo and engine. With a stock exhaust manifold on a DSM, there is only limited space between the manifold and block. Compressor housings that are much larger than stock will require you to dent the main water line that runs between the water pump and the left side of the engine. If you run into interference, dent the pipe slowly and gradually so that it doesn't crack. There is always that possibility, so be prepared to pull it if a crack or leak develops in the often-rusty pipe.

Most readers will probably be reading this chapter with the hope of replacing a perfectly good turbo with one that will make more power. If you're replacing a bad turbo with a new one, however, it's a good idea to know why the original turbo failed so

Same goes for the manifold-head nuts. Buy new ones before you start. Always use the OEM copper-coated nuts because they are much less likely to seize onto the exhaust studs than standard cad-plated or plain nuts. They're expensive, but worth it.

you can prevent the same thing from happening to the new (and probably rather expensive) turbo.

Did it start smoking from bad seals? Wear? A dirty turbo oil supply will destroy your new turbo very quickly, so verify that you have a clean, new oil feed line to the block. Make sure the oil drain line is clear and not kinked in any way. If you're having any head work done at the same time, make sure you have the oil galleries spotlessly clean so that no blasting media can get into the turbo oil supply.

Always perform a boost leak test after any turbo replacement. Fix any boost leaks, and carefully check the crankcase ventilation system to make sure it is working properly. Replace the PCV valve if necessary. Finally, pre-oil your new turbo carefully by cranking the engine over with the plug wires removed until the oil light goes out. Then break it in with low (wastegate only) boost for a few miles. Once you're sure everything is in good shape, feel free to turn up the boost and tune your new setup.

Intercooler Upgrades

While you might be tempted to consider an intercooler upgrade a secondary or later mod in your search for power, it might be better thought of as a crucial part of your overall turbo system. There are two reasons for this. First, as the airflow through an intercooler and its plumbing increases, the pressure drop also rises. At 250 cfm you might see a drop of only .5 psi or so. But near redline and running 500 cfm, your pressure drop could be 2.0 psi. A larger turbo that's able to push more air will very quickly tax a stock intercooler.

Second, a larger and more efficient intercooler will have more thermal mass that will allow you to push a turbo beyond its most efficient range. It's not the optimal solution to a mismatched turbo and engine, but it is a very good way to get low-end response from a small turbo without taking such a hit on top, at least during short bursts of high pressure. A good autocross setup might be something along these lines: a small,

The upper coolant feed line (on the left) should always be replaced if you don't know when it was last done. A failed hose can result in an overheated engine and junk head. The turbo oil feed line (on the right, since this is a 1g motor) should be inspected or replaced too. If you install a different turbo, chances are the stock hard line won't reach. Don't bend it —just replace with a flexible braided line. Be prepared to weld up the takeoff if the banjo bolt breaks off.

The Evo coolant lines are a little simplified from the DSM system. They're also easier to get to, since they're on the right side of the turbo (facing the engine). Chances are most Evos still have the stock hoses, but give them a once-over for cracks or tearing.

high-flowing turbo like a 14B matched up with a large front-mount intercooler.

If you have a big turbo running more boost, the temperature of the air blowing into the cooler might not be very high, but it will be hot for a long time as the turbo keeps boost up near redline. This will quickly heat soak a small intercooler. Your stock intercooler is designed with more thought given to packaging and intake ducting concerns than absolute efficiency. Unfortunately, that means it's usually pretty small and will heat soak (get inefficiently hot) in a hurry. If you've ever seen a turbo car on the dyno, you'll recognize the effects—big power on the first run with decreasing numbers on each successive pull.

The Galant VR-4 has the most restrictive intercooler and plumbing, so it should be among the first parts upgraded in the hunt for more power. Don't bother with changing the intercooler pipes until you change the intercooler. Unfortunately there aren't very many choices for bolt-in Galant intercoolers. Any intercooler kit you go with will likely require moving the battery and changing your intake.

On a DSM, you can increase cooling airflow to the stock intercooler by cutting holes in the fender liner, which will help somewhat with reducing heat soak. It won't do anything for the pressure drop across the stock intercooler, but you won't really need to upgrade until you do a turbo swap to something as large as the 16G.

You can reduce the pressure drop across the stock intercooler by modifying the inlets and outlets for larger hoses and more direct connection, but this will also require a change in intercooler piping.

Some aftermarket intercoolers for the DSM cars feature simplified plumbing that requires a rotated turbo compressor housing, oriented so that it is "downward firing" instead of blowing through the stock J-pipe (1g) or turbo outlet bend (2g). The easiest intercooler to install is the Greddy unit for the 2g, but it's not cheap.

When you choose an intercooler, think more about the manufacturer's reputation than anything else. Most of them will produce power within a few percent of each other. The biggest advantages they have over stock are the bar-and-plate core, better plumbing and end tanks, and greater thermal mass.

It's hard to go too big on an intercooler, especially if you plan to upgrade your turbo later. Large intercooler volumes can make for extra delay between mashing your foot and seeing boost, and are harder to mount behind the bumper. Most of the larger intercoolers require removal of the stock fog lights.

Replace the copper crush washers under each of these banjo bolts if you have to disturb them. Many replacement turbos come with water lines already installed, so don't be surprised if you don't have to do anything to them.

The stock oil drain line is known for cracking right at the bellows. If you have an oil leak from the front of the engine take a good look at the drain. Flexible silicone or rubber drains are available, but there is nothing wrong with the stock one if you're running a stock or stock-replacement turbo.

The stock Evo VIII and IX intercooler is big, free flowing, and right in the air stream. Anything less than 325 hp should be fine with the stock intercooler, but if you're planning to push the limits of the stock long block look into a replacement front-mounted bar and plate intercooler like this one.

INTAKE AND EXHAUST:

MAKING THE TURBO'S LIFE EASIER

The engine and turbo are the center of the power production system but they are not the entire system. The plumbing that leads from the air cleaner to the turbo, from the turbo to the intercooler, and from the intercooler to the throttle body all has an effect on system airflow. The intake and exhaust are the "ends" of the engine's breathing, and the more free-flowing they are, the more efficient the total system will be.

It may seem repetitive here, but efficiency is key to more power, and there is a reason airflow going into and out of the engine should be as high as possible. Turbos operate on the principle of a pressure ratio; the higher the intake pressure, and the lower the exhaust pressure, the easier it will be for the turbo to produce compressed air from exhaust energy.

Atmospheric pressure isn't always the same as the turbo inlet pressure. A restriction before the turbo will decrease air pressure on the turbo side of the restriction. The harder the turbo sucks (the more boost it's producing), the lower the pressure gets. This is the air that is "seen" by the turbo, so it's used as the

Everything you can see under the hood of this clean 2g, from the open-element air cleaner, to the large MAS, cast-aluminum intake pipe, stainless tubular exhaust manifold, and massive aftermarket throttle body, exists for one purpose: to make the turbo's job easier. The greater the airflow into and out of the turbo, the more power can be made.

first part of the pressure ratio (PR) calculation. The turbo seeing lower pressure raises the pressure ratio beyond what you would expect, based on boost and ambient temperature and pressure, which can make the turbo run less efficiently than you expect.

Restrictive boost plumbing to the manifold and head will also increase the pressure ratio. If the boost measured in the manifold is 15 psi, and the intercooler and plumbing cause a 1.0-psi drop, the turbo is actually producing 16 psi at the current airflow.

However, because the compressed air is denser than outside air, outlet plumbing restrictions have a much smaller impact on pressure drop than inlet plumbing restrictions.

If there are massive pre- and post-turbo restrictions, the actual pressure ratio can be much higher than you might think. This could put the turbo into a lower efficiency flow area, heating up the air more than expected. Conversely, a stock turbo in stock boost setup can be made to produce more power than stock as long as air can flow more easily before and after the turbo. As the compressor becomes a flow restriction, opening up the intake tract on either side of it will make no difference in power. Generally in this case, boost will start to fall off on the top of the rev range.

The exhaust turbine works a lot like the compressor, but in reverse. It has its own pressure ratio between the pressure in the exhaust manifold and the pressure in the exhaust system. It's similar to the compressor pressure ratio, in that changes on the low-pressure side (the exhaust system in this case) have more of an effect than changes to the high-pressure side (the manifold). A ported exhaust manifold will help increase the pressure seen by the turbine, but since the exhaust gasses are under pressure, it won't result in a large change in power output. On the other hand, a large, free-flowing exhaust will lower pressure on the output side of the turbine, which further reduces the pressure ratio.

Concentrate on the low-pressure side of the compressor and turbine more than the high-pressure side. Always go for the point of most resistance. The old strategy of upgrading your MAS and downpipe before anything else has its basis in science.

For example, say your engine and turbo combination is pushing 14.7 psi (gauge pressure, or 29.4 psi absolute pressure) into the intake manifold. The actual gauge pressure produced by the turbo is higher than 14.7 psi thanks to the restriction of the intercooler and intake plumbing. For the sake of illustration, we'll say it adds .5 psi for an actual turbo outlet pressure of 15.2-psi gauge pressure.

On the intake side of the turbo, the pressure is actually lower than the ambient (outside) pressure because of the restriction of the MAS and air filter, as well as the plumbing. Instead of the expected 14.7 psi (absolute) at sea level, the turbo sees less; for illustration let's say .5 psi for an actual intake pressure of 14.2 psi absolute.

Both restrictions change the pressure ratio created by the turbo compressor. This has an effect on turbo efficiency, as seen in the last chapter. The boost pressure ratio of this setup is 2.10 (14.7[atmosphere] + 15.2 [comp outlet] / 14.2[comp inlet]).

To show the effects of modification, notice what happens when you change the intercooler to eliminate the outlet pressure drop (keeping the same boost level). Without changing the MAS, your pressure ratio goes to 2.07 from 2.10 (14.7[atmosphere] + 14.7[comp outlet] / 14.2[comp inlet]).

The turbo is not working as hard as with the stock setup, but removing the intake restriction (and keeping the intercooler) makes a bigger difference. The pressure ratio drops to a much-lower 2.03 (14.7+15.2/14.7). So the same pressure drop (.5 psi in this case) is much worse on the low-pressure side of the compressor than on the high-pressure side. The same relationship holds for the exhaust: Exhaust-side restrictions are much worse than engine-side restrictions. This principle should make it clear where to spend your money first.

Pre-Turbo Airflow

The simplest inlet modifications are an open-air filter or cold-air intake (detailed in the bolt-ons chapter), but there are more modifications

The stock DSM intercooler (and Galant VR-4) is quite restrictive and doesn't get as much airflow, but it could mounted in the nose of the car. That said, don't feel the need to swap it out right away—the gains from an intercooler swap won't be as noticeable as power increases from improved turbo inlet breathing.

An open-element air filter is the first step in opening up the engine's airflow. A large, free-flowing element makes the turbo's job easier by increasing the pressure "seen" by the compressor. Stick with good-quality elements from known manufacturers to make sure you're getting a good part that will last longer than a few weeks.

The stock Galant VR-4 and 1g MAS and intake plumbing is restrictive. Notice how small the tube between the MAS and turbo is, as well as the small diameter of the MAS outlet. A swap to a 2g or Evo MAS will provide less restriction and allow measurement of more intake airflow.

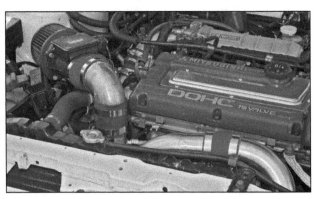

The outlet on the larger 2g MAS is oval, not round like the 1g MAS. The simplest way to mount any larger MAS is with a MAS adapter (you can find cheap ones on eBay), a good-quality cotton-element air filter, and an aftermarket turbo inlet tube like this one.

The larger Evo MAS is also shared by a few other Mitsubishi and Chrysler vehicles from the late 1990s, though with a slightly different bellmouth inlet. Any of these MASs will work fine on a tuned 4G36t.

you can make. After removing the filter housing or replacing the whole thing with an open intake, start looking at the MAS and the intake-to-turbo plumbing.

If you're running a bigger turbo you will have to tune the engine, so you might as well swap to a larger MAS. If you have a way to calibrate the ECU, swapping the MAS is an easy way to pick up a few horsepower and faster spool. The 2g or Evo MAS is an easy swap for a 1g, but don't forget to tune for it.

In Chapter 6 we'll get into more detail on engine management, but the 2g DSM MAS is larger than the 1g DSM MAS, and can measure airflow as much as 25 percent more than the 1g MAS.

The next step up is the Evo VIII and IX MAS, which is larger than any DSM MAS. It can measure as much as 50 percent more air.

If a stock 2g inlet tube won't work for you (i.e., if you have a 1g or VR-4) you'll have to fabricate your own tube to use one of these MAS units. If you can't find an off-the-shelf solution, you can make one with creative use of exhaust mandrel bends, silicone couplers, and the remainder of your stock tube.

The ultimate way to eliminate the restriction is to use a different sensor, such as a MAF sensor located after the turbo, or a MAP sensor with a different ECU. Any MAS change will require tuning—read Chapter 6 for more details on what has to be done to use a different or modified MAS.

Turbo / Intercooler / Intake Plumbing

On a DSM, the charge air bends out of the turbo, does a 120-degree turn into the lower intercooler pipe and flows out of the engine compartment to the intercooler. From the intercooler it runs around the right side of the engine to the throttle body and intake manifold, with the BPV attached before the throttle body.

The Evo is a little better in routing and tube size, but the massive length of the plumbing increases lag and doesn't do anything for flow. The lower pipe is relatively short and direct, but it necks down severely as it passes behind the fan. The upper pipe is also very restrictive as it snakes over the left side of the engine and competes with the intake tube for limited under hood space. Both pipes can be replaced for better high-rpm airflow and more high-boost horsepower.

The stock pipes on 4G36t cars are made of plastic and rubber, and quite restrictive. They're easily upgraded in the search for more power—even stock turbo setups will benefit from aftermarket intercooler pipes. You might also consider replacing the BPV if you have a 2g, and of course don't spend money on intercooler plumbing for a Galant VR-4 without replacing the intercooler.

The best (and possibly cheapest, depending on your fabrication abilities) intercooler pipes are custom-bent and welded for your particular setup.

MAS Sensor Plug Wiring

All Mitsubishi Karmann-Vortex MAS sensors are similar electronically. They can be swapped from car to car, but you will have to modify your existing harness connector to use a different MAS. Many late-model Mitsubishi and Chrysler cars use the same MAS connector, so you shouldn't have trouble finding an alternate connector at the junkyard.

Whatever you do, make sure that your wiring is connected properly before powering-on your new MAS—they can be damaged if the main power supply is connected to the wrong pin.

Note that these diagrams are looking into the back of the harness connector or the front of the MAS connector; reverse the wires if you're looking into the front of the harness connector.

1g:

1 GY	2 GL	3 R	4 GR
5 NONE	6 GB	7 GY	8 GO

1 - Idle position switch
2 - Airflow sensor
3 - MPI control relay
4 - 5V power supply for sensors
5 - None
6 - Ground for sensors
7 - Barometric pressure sensor
8 - Intake air temp sensor

The 1g DSM cars and Galant VR-4 have the same 8-pin, 2-row connector. The wire colors differ slightly between the two engines, though the sensors are interchangeable. Most of the time you'll be cutting one of these connectors off to replace it with a 2g or Evo connector.

2g:

1	2	3	4	5	6	7	8
GY	O	LY	R	B	RL	RW	None

1 - 5V power supply for sensors
2 - Barometric pressure sensor
3 - Airflow sensor
4 - MPI control relay
5 - Ground for sensors
6 - Intake air temp sensor
7 - Volume airflow sensor reset signal (Same function as "idle position switch" in 1g manual)
8 - None

The 2g MAS connector has eight positions in a row, although only seven of the positions are filled by a terminal. You need one of these connectors to convert your 1g to a 2g MAS, even if you build an adapter harness.

Evo:

1	2	3	4	5	6	7	8
GR	YL	WR	RY	B	RL	LB	None

1 - 5V power supply for sensors
2 - Barometric pressure sensor
3 - Airflow sensor
4 - MFI control relay
5 - Ground for sensors
6 - Intake air temp sensor
7 - Volume airflow sensor reset signal (Same function as "idle position switch" in 1g manual)
8 - None

The Evo MAS connector has the same seven wires as the 2g MAS connector, but the connector body is a different size and shape. It's very easy to wire one of these MAS sensors up to a 1g or 2g wiring harness. Make sure you have a way to modify the ECU programming if you want to run one, however.

Color Codes:
GR = Green/Red Stripe
YL = Yellow/Light Blue Stripe
WR = White/Red Stripe

RY = Red/Yellow Stripe
B = Black
RL = Red/Blue Stripe
LB = Blue/Black

Generally there is power to be gained from better intercooler piping with the stock intercooler and a slightly larger turbo, so don't feel like you have to upgrade the intercooler before replacing the pipes.

The Evo's upper intercooler pipe is not perfect, so a larger aftermarket unit like this one can help performance at high RPM with lots of boost and a larger-than-stock turbo. The benefits are debatable on a stock-turbo car, but if you've done everything else and don't want to swap turbos, it might be a good idea.

The upper pipes are the worst, and should be replaced first. The most horsepower-per-dollar gain will be the upper pipe (shown here in red). It eliminates several stock pieces including the restrictive throttle body elbow. The lower pipe can give significant gains as well.

Study this picture closely if you want to create your own custom intercooler plumbing. Notice the clean, tight welds and accurate bends. This shows the hallmarks of a professional tuner; in fact, it was fabricated by a premier Japanese shop from lightweight titanium and aluminum-alloy tubes.

You can use simple aluminum or mild-steel mandrel bends to build up any plumbing you can think of and tack them together as you go along. Even if you don't have a welder, it's possible to fabricate your own pipes with short silicone couplers.

One thing to keep in mind is the possibility of the piping blowing off.

Bead the ends if you have access to a tubing bender. If not, you can weld a couple of short beads at on the ends of the tubes. The short welds will be as good as a full bead around the tube. Be sure to wire-brush or grind the short welds down to keep the hose from tearing.

If your budget doesn't stretch to silicone tubing, you can use pieces of fabric/rubber radiator hose. The best place to find the good stuff is a semi-

truck parts or repair shop. They stock it in inside diameters from 1 to 4 inches. It's not cheap, but it does cost less than silicone.

For clean, straight cuts in your silicone tube sections, use hose clamps lightly tightened around the tubing as a knife guide. Use very sharp razor blades and try to keep the blade parallel to the surface of the tube at all times. Practice on a piece of scrap before you start cutting

Bead the ends of all home-fabricated intake piping to prevent the hoses from blowing apart under boost. A large boost leak like that will cause the engine to run massively rich. Short weld beads will achieve the same result, but they won't look as nice.

Silicone elbows make fabrication of some plumbing much easier. Again, buy quality stuff so that it doesn't blow apart under boost or tear when you're tightening it down. Quality elbows in the required diameter can be expensive, but it is resistant to high temperatures and will last the life of the car.

up your expensive length of silicone tubing. Trim your tubing gradually, 1/16 inch at a time, to avoid cutting it too short.

Some of the most important parts of the intake system are the hose clamps. The stock hose clamps (and their parts-store replacements) aren't bad, but they are narrow, and the inside of the slotted portion of the band is sharp. The stock narrow

The best way to connect pieces of hard intake plumbing is short silicone couplers cut from longer pieces of silicone tubing. Buy quality silicone tube since it will have better resistance to blowing off the ends of the hard tubes. It's available in just about any color you need.

bands won't hold onto slippery silicone hose for very long under high boost pressures and tend to let the hoses blow off.

Not only that, as you tighten the stock clamps onto a silicone tube, the threaded slots will grate the outside of the tube like a cheese grater. After a few times of this, your expensive silicone hose will start to look pretty bad.

The ultimate hose clamps are stainless "T-bar" clamps. They don't have as wide a clamping range as regular hose clamps, but the inside of the band is perfectly smooth, and the clamps can be tightened with a socket instead of a screwdriver.

These clamps are so strong they can even collapse hard intake tubes if you aren't careful. There are a few different types. The best to use on silicone are the constant-torque clamps, so-called because a small tight coil spring maintains a constant level of clamping force on the hose. These clamps also come in narrow and wide varieties for even more clamping force.

The budget alternative to fancy T-bar clamps are extra-wide, heavy-duty hose clamps. These are similar to your stock threaded hose clamps,

T-bar clamps (on the left) are the best hose clamps you can buy. Notice how wide the band is, and how the edges of the band are radiused to protect the outside of your connecting hoses. Heavy-duty hose clamps (middle) are an excellent everyday alternative. Don't use standard hose clamps (on the right) because they will tear up the outside of your hose, and can blow off under high boost.

Catch cans are a good idea for any turbo car, especially one with a few miles on the odometer. This is the simplest kind—it's vented to the atmosphere from the valve cover. A more sophisticated method uses two catch cans and a one-way check valve to allow the engine to burn the crankcase gasses when there's no boost. You don't want oil in the combustion chamber under boost.

but they have smooth, wide bands and large-diameter hex heads to tighten them with a socket. These clamps are also available from your local big truck parts store or race

The small breather hoses all over your engine bay can also be replaced by silicone hose. It will last almost forever. Try to cut it more cleanly than shown here for a sanitary installation.

Nylon cable ties are the best way to prevent small hoses from blowing off. The bad thing about using cable ties as hose clamps is removing them—they have to be cut off, and usually the only way to get to them is with angle cutters.

shop. If you can't get them locally you can find them at Road/Race Engineering.

While on the subject of intake plumbing and hose clamps, don't neglect the small vacuum hoses that run to the EGR valve, the charcoal canister, etc. Turbo engines are hard on underhood rubber parts, so replace them as regular maintenance. Some of the smaller molded hoses cost a fortune to replace with OEM Mitsubishi parts. Most can be replaced with straight hose from the parts store. Try to find and use a good brand.

High boost can actually blow these small hoses right off their nipples, so use proper-sized hose, and small hose clamps. Use either small spring clamps (sometimes called Corbin clamps) or nylon cable ties. Cable ties are cheap and fast, but the cut end is a hazard to the skin on the back of your hands when you're working on your car. A soldering iron is a good way to melt the end slightly and remove the sharp edge.

Throttle Body

The throttle body determines the amount of air that enters the engine at all times. The larger the diameter of the throttle body, the larger the amount of air that can flow through it. Assuming the engine can use more air—not always a good assumption—a larger throttle body will allow it to make more power.

The Evo VIII/IX and 1g DSM, have a throttle body that is big enough already for the engine's airflow, and a larger one will make little or no difference, since the engine cannot use more air. A large throttle body can even be a bad thing; the larger the throttle body, the less control the throttle plate has over airflow into the engine. A throttle body that is too large will make the engine jumpy and hard to drive smoothly.

If you want to try a larger throttle body, you have a couple of options. The first is to port the stock one. While this does not usually increase the size of the throttle plate, it does smooth airflow into and out of the throttle body, and can increase total airflow when the throttle is wide open. Several companies offer this service since it can be a little more than a do it yourself project.

The second choice, if you have a 2g, is to use the throttle body from a 1g. The throttle you want is from a 1g car, with four vacuum connections on the top facing the left side of the car. The earlier one (also with four vacuum connections, but two on each side) won't work. You reuse the 2g TPS (the 1g TPS has an extra switch for closed throttle inside).

The vacuum lines on the 1g throttle body require a little reworking to use with the 2g's vacuum system. Mate up the matching codes with their matching line, except for

The Evo VIII and IX have a large throttle body with a bead around it instead of the restrictive cast elbow found in front of a 1g or 2g throttle body. The Evo throttle body is usually not replaced until power levels get very high.

The 60-mm throttle plate of the 1g throttle body (on the right) is approximately 33 percent larger in area and flow than the 52-mm 2g throttle body (28.30 square centimeters versus 21.22 square centimeters). The stock 2g throttle body isn't a restriction until you change the turbo and intercooler, though.

It doesn't take much porting to open the 2g intake manifold out for the 1g throttle body. Go slow and you shouldn't have a problem. Don't worry about blending the shape deep into the intake manifold; it's just not necessary.

the EGR line, which should be tee'd off the line from the "P" port on the throttle body so that it gets the proper constant vacuum signal. Block the stock 1g EGR line (marked "E").

Don't forget to port the 2g intake manifold to eliminate the step that would otherwise exist (and defeat the purpose of the larger throttle body). This can be done with the manifold on the car and assembled; just pack it full of shop rags before grinding away on the manifold. Use the 1g throttle body gasket as a template, and scribe a porting line around the inside diameter of the gasket.

Grind with your favorite porting tools. Use a shop vac to remove as much of the chips as possible before removing the rags stuffed in the intake manifold and do everything you can to get the chips out before bolting on the new throttle body. Adjust and reset the throttle position sensor (TPS) and you're almost done.

The procedure for adjusting the TPS depends on the type. For a 2g, adjust by measuring for continuity between pins 3 and 4 on the connector (the idle switch pins). Make sure that the fixed throttle stop is clear of the throttle arm by .018 inch first.

Turn the TPS as far to the left (counterclockwise) as you can with the screws loosened, then turn slowly back to the right (clockwise) until the idle switch opens (you measure no continuity between pins 3 and 4).

For both 1g and 2g TPS, double-check the adjustment by measuring the resistance between pins 2 and 4 on the connector; it should be around 1k ohms. Mitsubishi specifies a voltage level of 500 millivolts between pins 2 and 4 with the ECU on, but it's harder to measure the voltage without a jumper harness.

The final step in installing a larger throttle body is the elbow that connects to the intercooler plumbing. You can either use the matching 1g elbow for a 1g throttle body, or port the 2g elbow to match. Don't get too crazy; just blend the diameter in a little bit and get rid of the step inside. The casting of the elbow is very thin and you can easily break through to the outside if you're not careful.

If a 1g throttle body isn't big enough, there is a 62-mm replacement available from BBK. It doesn't

The 1g throttle body elbow is the most restrictive of the DSM elbows. The 2g elbow, though it has a smaller outlet, is larger throughout than the 1g elbow. You're better off porting your stock 2g elbow if you do a 1g throttle body swap. Evos, luckily, don't have an elbow to restrict airflow through the intake plumbing.

require all of the modifications needed for a 1g throttle body swap, but it's more expensive. You can also have a stock throttle body bored to around 58 mm, or have a 1g throttle body bored to 63 mm.

Because of the amount of work this conversion takes, don't do it until you've swapped turbos and replaced your intercooler plumbing, at a minimum. Even the stock throttle body flows more air than the stock intercooler and plumbing.

Aftermarket adapters and manifolds can be designed to adapt even larger throttle bodies, but the result is a throttle pedal that's jumpy and hard to modulate in normal traffic. Anything larger than 60 mm isn't really needed for less than 500 hp, so save them for the drag strip.

Intake Manifolds

The 1g manifold is arguably the best production DSM intake manifold. If you have a 1g, you won't see much improvement from swapping out the stock piece. The 2g manifold is

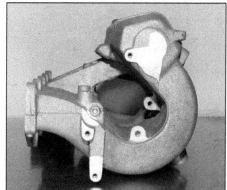

The stock 1g intake manifold is a good piece. It flows moderately well for a production car, and it has large, open ports. Notice how the ports shoot straight into the head. This is not the most efficient way to get air into the combustion chamber, since the floor of the port is so low. It works very well at high flow rates, though.

similarly good, but it has smaller runners feeding the intake ports in the head.

The name of the game in getting good midrange performance from an intake manifold is good flow and good velocity. Turns out, the 2g intake manifold runner size and shape is almost perfect for a street car. Swapping to a 1g manifold isn't really helpful, and in fact may hurt power production except at the very top of the rev range (if at all).

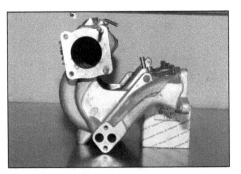

The 2g manifold has runners about the same length as the 1g manifold, but notice how they turn the air down toward the valves more. This increases velocity and midrange performance compared to the 1g manifold. It's a good street manifold.

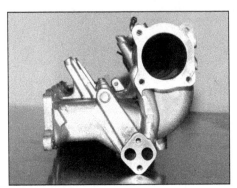

The Evo manifold is clearly a development of the 2g style of manifold. It has good, high-angled runners that allow it to flow very well, as well as provide good midrange air velocity. But notice how short they are compared to the DSM runners— this is a high-RPM, high-velocity intake manifold.

The best of the "regular" (right-facing) 4G36t manifolds is the one fitted to the Evo III. It's almost a bolt-on for your 2g engine and has the potential to flow much more air. It doesn't have a provision for an EGR valve, which is an issue for some people in smog-controlled states, but it works very well for street applications.

There isn't much to be gained from swapping the Evo VIII and IX manifold for anything else. Lots of

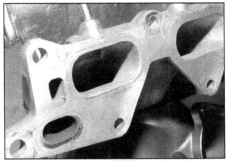

The stock manifold is a good piece, but if you want to squeeze a little more flow out of it you can clean up the ends of the ports. Don't remove too much material, and don't try to change the shape. This 2g manifold shows the kind of shape and finish you should aim for.

Most aftermarket intake manifolds (like the Magnus one shown here) are fabricated from sheets of aluminum alloy welded together. Others, like the HKS manifold for the Evo VIII and IX, are made of cast aluminum. Either one works fine, but fabricated manifolds tend to be a little cheaper.

cars are making 400–500 hp at the wheels with the stock manifold. Like most other 4G36t parts, you can't swap intake manifolds between Evo and DSMs, unfortunately, because of the engine's flipped mounting.

For really high-powered engines, like those reaching for the 500-hp mark (and 30+ psi of boost), aftermarket intake manifolds are the way to go. These incorporate larger plenums than the stock manifold, larger runners, and a more direct path for airflow into the intake port.

The best aftermarket intake manifolds have big, straight runners with a carefully radiused entrance inside the intake plenum. They also have a plenum roughly the same capacity as the engine's displacement, and are designed to fit within the confines of the stock engine compartment. The larger plenum allows each cylinder to draw from the air more completely, equalizing flow between cylinders, while the radiused runner entries give air the best shot at making it to each cylinder quickly.

In general, aftermarket intake manifolds help a little at the very

Notice how straight the runners are in the aftermarket manifold. They give air the best shot at the intake valves possible. The plenum is large, and the intake runners all have a carefully sculpted bellmouth inside the plenum. Don't buy an aftermarket intake without these features.

bottom of the rev range (below 3,000 rpm) by allowing the turbo to spool up faster, and more at the very top of the rev range (above 5,000 rpm). Testing by the Rocky Mountain DSM Club in 2004 showed fairly clearly that a stock 1g intake manifold will outflow even the best aftermarket manifolds in the 4,000 to 6,500 rpm range.

If your turbo airflow peaks at around this point, you might be better off with the stock manifold. On the other hand, if you're going for all-out, high-rpm power, swap to an aftermarket manifold and don't look back.

Insulating Spacers

A relatively new product on the aftermarket scene are plastic insulators designed to go between the cylinder head and intake manifold, reducing heat transfer from the hot head to the manifold. These work by lowering the temperature of the intake manifold, which increases the density of the air that reaches the valves.

The spacers limit temperature rise time, which is a good thing, but they do not do much for ultimate heat-soak temperatures. Your intake charge is so hot, even after the intercooler, that eventually the manifold will reach its thermal capacity regardless of whether it's connected directly to the cylinder head or insulated.

The spacer will help the intake cool off quicker, too, which is another benefit for street drivers and drag racers. Road racers might not notice much change in intake manifold, and thus charge temperatures, since the constant-rate temperature of the manifold will not change much.

That said, intake spacers don't have much of a disadvantage other than cost and the difficulty of installing them. They're worth a lit-tle power but the price-per-horse-power ratio probably isn't very good.

Exhaust System

Turbo engines love a free-flowing exhaust system. The Mitsubishi turbo exhaust system as delivered is pretty restrictive and not close to what the engine really wants. Exhaust blasting out of the turbo wants to be cooled and evacuated as quickly as possible to convert more of the exhaust heat and pressure into energy. The gasses flowing out of the turbo are also turbulent; a narrow restrictive exhaust causes the turbulence to bounce back into the exhaust stream and hurts flow.

From the cylinder head, exhaust collects together and flows into the turbine housing. After it gives up its heat and energy for the benefit of your right foot, it flows out of the turbine and wastegate together to the O_2 sensor housing (an elbow that directs the flow back downwards). After the O_2 housing, it dumps into the downpipe, and from there it proceeds to the catalytic converters and rest of the exhaust.

Each of these parts can put their own restriction onto the gasses flowing into and out of the turbo. If you want to quantify each restriction you could use a pressure gauge connected to various ports in the exhaust system. A gauge in the O_2 housing will show the pressure in the entire system, and you can record the difference as you change parts around.

Exhaust Manifolds and Headers

All DSM and Evo models have a similar cast-iron exhaust manifold. The 1g automatic manifold is the smallest of the stock USA-market manifolds, with a smaller opening and turbine bolt pattern for the TD04h turbo these cars were equipped with. The manual 1g and Galant VR-4 manifolds are a little larger in the turbine opening area, but they are not particularly free-flowing and they are very prone to cracking.

The 2g manifold is the best of the DSM manifolds. It flows fairly well, has the TD05h bolt pattern, and does not crack as often as the earlier manifolds. It is almost a bolt-on to the 1g engine; the only differences are a couple of 8-mm bolt holes (versus 6 mm on the 1g) and a little bit of interference between the manifold flange and the power steering bracket on the 1g head. The first problem requires a couple of thick washers, and the second one is easy to resolve with a little grinding.

The best of the standard 4G36t manifolds is the one that came stock on the most powerful variant of this engine; the Evo III. This manifold has a thicker flange and thicker casting walls than the earlier manifolds, and is made of a better grade of cast iron. It never cracks, and the larger internal passages flow more exhaust than any of the stock DSM manifolds. It

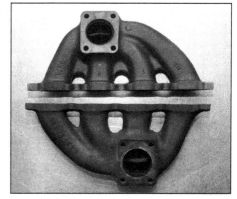

Most any stock 1g manifold (bottom) has cracked by now, but there are other reasons to upgrade to the Evo III manifold (top). Not only is it thicker, it also flows better since it has a larger collector that matches the larger 16G turbos. The Mitsubishi part number is MR224509.

The stock Evo VIII and IX manifold is a good part with plenty of potential for the stock twin-scroll turbo. Unfortunately, it cannot be used on any of the right-facing 4G36t engines because of turbo to engine interference.

For a cheap manifold upgrade, just port your existing one. A long as it's uncracked, a little port work is a good way to get more flow without spending a lot of money. Good results can be obtained from a port job like this one; don't go all the way to the gasket outline because the gasket is larger than necessary.

also has a turbine housing flange sized to fit the larger Evo III 16G turbo's exhaust housing. It bolts on the same way as the 2g manifold.

The Evo VIII and IX manifold is, of course, rotated to match that engine's reverse-spinning turbo and flipped head. It also has an internal divider that keeps the exhaust flow from the number-1 and number-3 cylinders separate from the exhaust from the number-2 and number-4 cylinders. This matches the split inlet of the twin-scroll turbo.

Manifold Swapping and Porting

If you have a 1g, you probably don't have the stock manifold because of cracking. If you do, replace it with a used 2g manifold for the cheapest possible upgrade, or a new Evo III or aftermarket piece. For more flow, all DSMs can benefit from the Evo III manifold, especially with a larger bolt-on turbo like the Evo III 16G.

The Evo VIII and IX manifold, on the other hand, is fine for pump-gas power goals. On all turbo engines the engine side of the turbine is not nearly as important as the exhaust side of the

turbine. Unless you have a turbo at least as large as a big 16G, your turbine housing will be a much bigger restriction than the exhaust manifold.

If you do the work yourself, porting the exhaust manifold is a good way to get more flow with a larger turbo. Don't go too big, just grind the manifold face out to match the exhaust gasket for the engine you're working on. This will result in a small "step" between the port and mani-

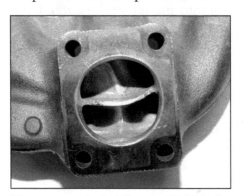

It's more work, but the manifold-turbo flange on a DSM manifold is easily ported out to remove the step that usually seals against the sealing ring. Ditch the ring and use a flat manifold after porting. Your gasket might blow out sooner, but flow will be increased.

fold, which is good because it adds a little anti-reversion to the port. Anti-reversion helps the most at part-throttle and low boost, but it only hurts flow at very high flow rates (such as with a giant turbo and lots of boost).

All of the DSM and early Evo manifolds had a large sealing ring between the turbo and manifold that fits into a step in the manifold. The Evo VIII and IX manifolds do not have this step. Grinding out the step on the manifold and the turbine housing is a good way to pick up some additional airflow from the stock manifold.

Aftermarket Manifolds and Headers

If you have an Evo, your journey for more exhaust flow starts with the aftermarket. The stock manifold is very good, however, so don't jump to a new one unless your power goals are stratospheric. An aftermarket manifold on a turbo car will tend to increase lag slightly, since the larger volume of the big runners takes longer to fill and pressurize with exhaust gasses. If you're going for big power numbers or low drag times, the extra high-boost flow will

Cast-iron manifolds are heavy and thick, which makes them less beautiful than tubular steel manifold. However, they are much preferred for street applications because they are more durable than tubular steel, and their greater thermal mass helps them keep heat in the exhaust and spool the turbo faster.

outweigh any losses at part throttle and low boost.

The DSM cars perform similarly once they've been upgraded to an Evo III manifold. A ported Evo III will flow enough for a 350- to 400-hp street car. If that's not enough for your race car, there are lots of aftermarket options. On any 4G36t, if you want to run a big aftermarket turbo, you'll have to swap manifolds to get the right turbine bolt pattern anyway.

Aftermarket manifolds are made either from cast iron, like the stock parts, or fabricated from tubes of steel or stainless steel. For most applications, including all street applications, a cast-iron manifold is a better choice. The thick, heavy walls resist cracking, and they give you room to port the manifold to match different heads and turbine housings. The thick material also holds in heat better; and since on a turbo engine heat = power, they help spool and turbine performance.

In a tubular header, look for quality welds, good, thick steel tubes, and a professional finish. A cheap header will crack, flake, warp, leak, and generally turn out to be a waste of money. Be warned: turbo engines are

Exhaust manifolds for naturally aspirated cars are carefully designed to equalize the length of each runner, but turbo manifolds are usually not. An equal-length manifold can help, however, by equalizing exhaust scavenging between cylinders. In turn, this can help high-RPM exhaust flow and power.

very hard on headers. The heat on the engine side of a turbo is extreme, and it is hard for any header to handle the temperatures encountered. Multiple heating and cooling cycles stress the thin steel of a tubular header much more than the thick iron of a cast manifold. For this reason, many eBay-type tubular manifolds don't last long.

O$_2$ Sensor Housing and Wastegate Plumbing

After the exhaust gasses have been through the turbine housing, they've given up most of their pressure, heat, and energy. The gasses are still hot and pressurized, but the heat and pressure are much less than before the turbo. Airflow becomes even more important now, since the gasses are not as dense as before the turbo or in the intake manifold.

The 4G36t exhaust system starts at the turbo with a cast-iron bend called the O$_2$ housing because it's the location of the stock first oxygen sensor (O$_2$ sensor). The O$_2$ housing acts as a merger for the gasses flowing out

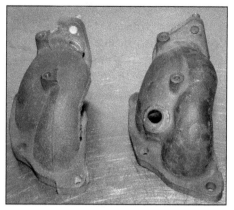

The 1g and 2g DSM O$_2$ housings are interchangeable. The 1g O$_2$ sensor port is near the top of the housing (right), while the 2g port (left) is near the bottom. The Evo IV through IX housing is not interchangeable with the DSM ones because of the turbo's orientation.

of the turbine and out of the wastegate opening to come together before dumping.

The DSM housing has some design problems that will work against you later. The area for exhaust to get out of the wastegate is too small, so boost can increase beyond the target even when the wastegate is full open (boost creep).

Of course the Evo III had the best O$_2$ sensor housing (on the right) of all the right-facing engines. The O$_2$ housing is larger and more free-flowing than the DSM and Galant part on the left. A ported DSM housing will never flow as well as the stock Evo III housing (part number MR224847).

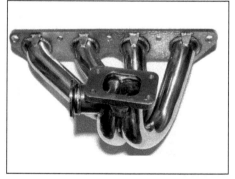

Manifolds welded up from tubes are the racers' choice. They are lighter than cast iron and offer more flow rate in the same area, since the runner walls are thinner. They are a lot more expensive and tend to be less durable, particularly on the street.

The Evo VIII and IX housings are better than the DSM housing, and boost creep is not normally an issue. A bigger issue is the small radius of the 90-degree bend in the housing and the pinched pathway for exhaust flow.

Aftermarket O_2 sensor housings are usually fabricated from stainless steel tubes. A higher-flowing O_2 housing will help a small turbo setup make more power by reducing after-turbo restrictions. Even more importantly, the higher wastegate flow and smoother transition to the main exhaust flow will reduce boost creep with large turbo setups on the street.

If you want to go really big, get an O_2 housing eliminator, which is basically a 3-inch diameter elbow that replaces the cast or aftermarket O_2 housing. You'll have to make sure you have the right downpipe flange, but it's the best way to get more flow in the stock layout.

As mentioned above, the stock wastegate outlet combines with the main turbine outlet flow in the O_2 sensor housing. This setup works, but the short distance before the merge affects turbine efficiency and spool, since the gasses from the wastegate cause turbulence that slows the main exhaust flow. You can improve on it with a small-diameter dump tube that parallels the turbine outlet pipe for a few feet and then merges into the downpipe under the car.

This allows the two streams to do their thing where it's most beneficial, and then flow smoothly together for the trip out the exhaust system. It can help with high-flow and low-boost (lots of wastegate opening) situations, but it won't have much benefit at any other time. It requires eliminating the stock O_2 sensor and flange, which isn't a bad thing, but it also requires some heavy fabrication. There are aftermarket downpipes

For the ultimate in boost control on a race application, leave the wastegate flow tube open to the atmosphere. It will be loud under full boost (really loud!), but there will be no issue with boost creep even with a wide-open 3-inch exhaust. This works for both internal and external wastegates.

designed with this in mind if you don't want to build your own.

There are bolt-on external wastegate setups available from the aftermarket for various turbos; check with your turbo or manifold maker for specifics. You can use an external wastegate with a bolt-on or stock turbo with a special O_2 housing/elbow that breaks out the stock wastegate port to a flange. To use an external wastegate with this elbow, remove the flapper door and actuator from the internal wastegate, and add an external one.

If you are running a big aftermarket turbo with a different flange, you have to use a manifold or turbine housing with provisions for an external wastegate. Most aftermarket manifolds and headers have a wastegate flange in the bolt pattern of your choice. Just add your wastegate, and, on the street, plumb the outlet into the downpipe before the rest of the exhaust system.

Exhaust Heat Control

Anything you can do to keep heat and pressure up in the manifold and down in the exhaust will give

The most basic form of heat control is the stock manifold heat shield. If you use an aftermarket manifold, you should make sure you have a heat shield as a bare minimum. It also has the side effect of keeping your fancy manifold and turbo out of the view of prying eyes who might be interested in your modifications for legal or other reasons.

you a boost in terms of spool time as well as ultimate horsepower. Turns out it's easier to keep the hot side hot, simply by insulating the manifold and turbine housing. Insulation will keep the heat in and prevent parts from cooling as quickly as they otherwise might.

Insulating the pre-turbo exhaust has other benefits as well. Keeping that heat in the exhaust has the side effect of preventing it from toasting your underhood wiring and plumbing. It also keeps heat away from air going into and out of the intercooler; that's air you want to stay as cool as possible.

Sometimes you can rework the stock heat shield to clear an aftermarket manifold, and sometimes you have to swap to an aftermarket shield. Either way will work fine, though the first is less expensive.

The next step up from the heat shield is glass-fiber exhaust wrap. Exhaust wrap can cut under hood temps in half. You can buy "blankets" that are essentially the same stuff but are pre-formed to go around the manifold, downpipe, or turbine housing.

Exhaust wrap keeps down under-hood temps and channels more heat energy to the turbo, but it encourages cracking and corrosion on a street-driven car. Race cars can get away with it, but there are more sophisticated methods for controlling under hood temps on the street.

All DSM and Evo downpipes are roughly similar, although there are small differences in the angle of the first bend and the length of both parts. For such a simple part, it has a big influence on horsepower. The gasses coming out of the O_2 housing are very turbulent; the better they can be merged into the rest of the exhaust the better.

Notice how this downpipe mates directly to the outlet of an aftermarket turbo. Instead of a restrictive cast O_2 housing, the exhaust gasses simply have to make a short right turn to exit through the downpipe. It's the best way to build a downpipe—it's even better if, like this one, it includes an external wastegate.

One downside to wrapping is that it tends to cause parts to crack, especially tubular exhaust headers. They're almost guaranteed to fall apart after a few years from cracking and corrosion, as the wrap traps moisture from condensation against the parts. Even cast-iron manifolds might suffer from heat wrapping, so consider the effects before starting to wrap everything in sight.

The most high-tech method of heat control is so-called heat barrier coatings. Mostly ceramic-based, these coatings are designed to reflect heat away from the coated part. They are designed to keep heat away from the substrate metal, which tends to help parts last longer since they are less likely to experience localized hot spots.

Downpipes

Continuing down the exhaust system from the O_2 sensor is the downpipe, which connects the turbo to the rest of the exhaust system. The stock DSM downpipe is a joke—a 2.25-inch OD pipe that's crush-bent to clear the oil pan and transmission. The Evo downpipe is a little better,

but it's still rather small for the engine's power potential, and crush-bent. Both downpipes are a similar shape—a sharp 90-degree bend before a relatively straight piece of pipe connecting to the catalytic converter and the rest of the exhaust system.

In general, the largest downpipe you can fit behind the turbo should be used. If you swap downpipes, consider a larger O_2 sensor housing at the same time. Some aftermarket downpipes have an integrated O_2 housing (or O_2 eliminators). These have the potential to flow much better than the stock-style setup, since they do-away with the stock flange that will always be the flow-limiting part of the downpipe/O_2 housing combination.

Exhaust System and Mufflers

The rest of the exhaust system from the downpipe back consists of a catalytic converter, resonator, piping, and muffler. If you're replacing a plugged-up, 15-year-old DSM exhaust with a brand new 3-inch cat-less system, you're going to see a bigger

improvement than someone swapping out their brand-new Evo's free-flowing stock exhaust. Unfortunately, though, cat-back systems alone can only do so much. The most restrictive parts of any exhaust system are the downpipe and catalytic converter.

Mufflers are not as important for power production as many people assume. Also, noise is not an indication of flow. Some of the best exhausts are pretty quiet, while there are very loud but inefficient exhausts available from the aftermarket, too.

There are two ways that a cat-back will improve your car other than performance: sound and weight. The stock system on your DSM weighs at least 70–100 pounds; a good aftermarket exhaust can shave off about half of that. Exhaust sound is more important to some people than others. There's nothing like a tough-sounding ex-haust to get you noticed, which is good in some situations but bad in others (like when that attention comes from the black and white on the corner).

Most quality aftermarket systems are made from mandrel-bent tubing. This means that each of the bends is done in a machine that does not reduce the diameter of the tubing at the bend. An exhaust system bent up by your local muffler shop will likely use crush-bent tubing, where each bend squeezes the inside diameter of the tubing by 1/2 inch or more. A mandrel-bent exhaust system can be made from smaller tubing without giving up any flow compared to a crush-bent system and is the performance choice.

Quality aftermarket exhausts are usually made from stainless steel for durability. Mild steel is probably fine if you live in a desert state, but even there internal condensation will eventually rust the pipes out.

When choosing an exhaust, the larger diameter the better from a purely performance standpoint. A 3-inch mandrel-bent exhaust is the standard for most performance street cars. A 2.5-inch mandrel-bent system is easier to squeeze under the car without rubbing, and the noise level is more street-friendly, however. If your power goals reach above 350 hp, go for the 3-incher. Lower power levels are not as sensitive to exhaust size, assuming the same downpipe and catalytic converter.

One good thing about turbo engines is that they require less muffling than non-turbo, since turbos tend to quiet exhaust pulses by absorbing their energy. If you go with really big pipes, be prepared for droning and knocking over bumps. As with everything else, a big exhaust is a compromise.

Muffler styles vary greatly between different exhaust systems, but for a performance turbo engine the best choice is some kind of straight through or turbo muffler. Don't use a chambered muffler with a turbo engine because the backpressure during high exhaust flow is unacceptable. Straight through mufflers have been improved recently, and can now be made very quiet.

A neat way to get around the compromise on a street/strip or street/track application is to add a "cutout" in the exhaust system. A cutout adds a "Y" to the downpipe before the rest of the exhaust system. A plate similar to the throttle plate closes off the open leg of the Y, with the exhaust connected to the other leg. By flipping a switch (electrical cutout) or pulling on a cable (mechanical cutout), you can open the other leg of the Y and have an open exhaust for the most power and noise. If you want to try a cutout, look for one that seals well, since any leak will lead to more noise than you might want.

Some people have rigged up their electrical cutout to open at full throttle, or after a certain PSI of manifold pressure, but there are drawbacks to running it this way. Running an open cutout on the street is an invitation for a ticket—any automatic system will need to have a manual override switch. Finally, many electrical cutout motors simply aren't designed for frequent use. If you use one all the time, be prepared for the motor to seize up after a while.

Emissions Equipment and Legality

With all the talk these days about global warming and air quality, it's up to us as enthusiasts to make sure that we don't draw too much unwanted attention in our direction. Driving a car that pollutes isn't just about you, it's also about the thousands of other tuners who would suffer if even more strict smog control laws were passed. On a more personal level, if you've ever lived in an area with poor air quality, you will have seen first-hand the results of ignoring the consequences of high exhaust emissions.

Emissions control laws vary by state, with states that follow the California Air Resources Board (CARB), having the most restrictive rules. In all of the United States, cat-back exhaust systems are legal on any car as long as they're quiet enough to pass noise regulations, but most other parts cannot be changed. CARB states place restrictions on changes to the header and other any part of the engine between the air filter and the last catalytic converter with a few exceptions, such as the intercooler, which can be modified freely, but none of the parts that connect to it can be.

If the manufacturer of your part spends the necessary money to prove to the CARB's satisfaction that the part does not change emissions, it may be granted an "EO" or Executive Order number that allows them to be installed on a smog-controlled vehicle. However, every state's rules are different, so always check your local laws if you are unsure of the emissions legality of a modification. These rules also change according to the political winds, so keep yourself educated on parts legality before modifying your car.

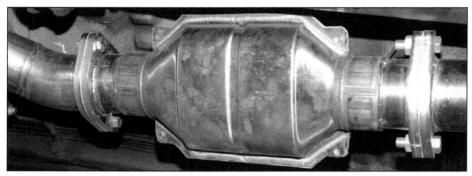

All modern cars, including all Mitsubishi cars, are equipped with one or more catalytic converters. There is no way around the fact that stock catalytic converters cost horsepower; even aftermarket cats like this one present some restriction. However, you should run one to stay on the right side of the law.

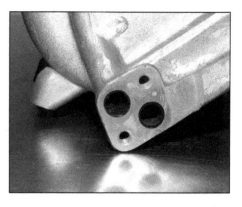

The EGR ports allow exhaust from a passage through the head to enter the intake. A lot of people block them off when they're tuning an engine, but there isn't much reason to. It reduces the weight of your car by maybe 1/2 pound, and it adds absolutely no horsepower.

Catalytic Converters

Catalytic converters are the single most important part on your car for maintaining low emissions. That said, they are a very big power restriction in your exhaust system, and you're almost guaranteed to pick up a few horsepower by removing them. A stock cat will cost you between 5 and 10 hp depending on the rest of your setup, but an aftermarket metal substrate hi-flow cat is less restrictive and will cut the power loss in half.

Removing catalytic converters from the exhaust will significantly increase the engine's emissions of smog-forming chemicals, including various unhealthy carbon compounds. The consequences are probably not worth it unless the car is a track-only machine with programmable engine management.

Removal of the converters will also put you on the wrong side of the law in the United States and many other countries. A cat-less car will definitely fail any kind of tailpipe emissions test, and if it has OBD-II engine controls, it will fail that portion of an emissions test as well.

Even a replacement catalytic converter is illegal in the USA in some situations. You can not replace a working catalytic converter with another one for any reason unless the car has more than 80,000 miles on it or is older than eight model years. In California, no catalytic converter may be replaced with an aftermarket equivalent on a car made after 1996. That means even a high-flow aftermarket cat cannot be installed in place of the original.

Exhaust Gas Recirculation (EGR) System

An EGR system is an emission-reducing device fitted to all USA-market DSMs and Evos. It recirculates a bit of the exhaust gas back into the intake manifold during cruise conditions to displace a little of the intake air and cool combustion temperatures. This reduces formation of oxides of nitrogen (NO_X), a significant contributor to smog. The EGR valve is operated by a solenoid controlled by the ECU; normally it's only active during cruise and not when the turbo is producing boost.

Having a functioning EGR makes it easier to get your car "smogged" if you ever plan to drive it on the street, and your neighbors will thank you for leaving it in since their air quality will be better.

There are a couple of situations where you will have to eliminate the EGR system. Some combinations of intake and exhaust manifolds prevent you from running an EGR (most aftermarket intake manifolds don't have a provision for it). The popular Evo III intake manifold has no EGR provisions either, so you will have to figure out some other way of adding EGR to your engine if you want to stay smog-legal.

The second situation where you will want to ditch the EGR is with an aftermarket engine management system. Most of these systems won't be able to control the EGR system efficiently; in that case it makes more sense to remove it. Of course you won't really need aftermarket engine management at street horsepower levels, so it's pretty much a non-issue.

FUEL SYSTEM

Once air is sucked into the intake tract, it must be combined with a precisely metered amount of fuel and ignited at just the right time to produce the controlled explosions that spin dyno rollers and launch us toward the end of the track. The engine's fuel system ensures that this all happens—from the tank to the engine, each part contributes to getting a carefully controlled dose of fuel (in this case gasoline) into the engine.

There are some facts about gasoline that you should know before attempting to tune an engine yourself, or talk to your tuner about what can be done. Gasoline burns best when it is combined with air at a ratio of 14.7 pounds of air per pound of gasoline.

This is known as a stoichiometric ratio. An air/fuel mixture in this ratio will burn completely with no unburned gas or excess air remaining. As mixtures get richer (a lower percentage of air by weight) the gas burns less efficiently and more unburned hydrocarbons are left behind. As the mixture gets leaner, it also burns less efficiently, but in this case it burns hotter.

In an internal combustion engine, the "perfect" stoichiometric

The fuel rail and injectors on this engine are ready to dump enough fuel to support more than 100-hp per cylinder. A free-flowing fuel system requires a lot of thought and planning to avoid an engine-melting mishap.

ratio of gasoline to air is not the best ratio for power production. It is the best mixture for exhaust emissions and fuel economy, however, since it wastes minimal fuel. For maximum power, air/fuel ratios should be in the range of 12.0:1 to 13.0:1. However, with a turbocharged engine, this may be too lean. Most tuners target 11.0:1

for a turbocharged engine since the cylinder pressures and temperatures are already high enough and the slight amount of extra fuel helps to cool them.

Lots of people are starting to experiment with E85, a fuel made of 85 percent gasoline and 15 percent Ethyl alcohol, but it is not widely available and requires a lot of tuning to extract the most power. The pay-off is there, though; E85 burns much slower than gasoline and therefore has a higher effective octane. The rich mixture that it requires (E85 contains less energy than gas for a given volume) also helps by cooling the intake charge, like water injection or an intercooler.

However, E85 is not a magic bullet for power. Since larger amounts of fuel have to be used to produce the same amount of power, it requires larger injectors (50–80 percent larger) and careful dyno tuning. You can't just add E85 to the tank and crank up the boost; the engine will then run dangerously lean. E85 also reduces fuel economy and can cause corroded fuel system plumbing if the parts you use are not compatible with Ethanol.

Octane

Octane is a measurement of fuel's resistance to knock, or detonation. In a nutshell, detonation is uncontrolled combustion. Instead of the air/fuel charge exploding in a controlled manner and pushing down on the piston evenly, it explodes awkwardly, sending shockwaves across the combustion chamber and hammering on the top of the piston. This produces the distinctive "ping" or "tink" sound that we learn to avoid at all costs.

In effect, detonation is like hammering on the tops of your pistons with a pickaxe. The mechanical damage from knocking destroys pistons, rings, head gaskets, cylinder liners, and sparkplugs. Severe detonation will actually melt chunks of aluminum from the top of the piston and pepper it around the inside of the combustion chamber and sparkplug.

The higher the cylinder pressure, the more catastrophic the effects of detonation. On a slow-revving industrial engine with heavy pistons, cylinder pressures are not very high and pinging will damage the engine slowly. In a turbocharged engine running a lot of boost and high RPM, cylinder pressure skyrockets to the point that even the tiniest amount of detonation can destroy the pistons. High-octane fuel reduces the chance of a spike in cylinder pressure by burning more slowly, and therefore preventing damage to expensive engine internals.

Ignition timing also plays a big part in the detonation story—too much timing advance encourages uncontrolled combustion and the resulting detonation. For a given boost pressure, RPM, and fuel octane rating, there is only a narrow range of about 4 degrees of timing between maximum power and knocking. This will become an issue later when we address tuning, but for now keep in mind that octane is one of the most critical variables affecting how much boost and timing (and therefore power) can be made with an engine.

Fuel Injectors

Fuel injectors are electrical valves that allow pressurized fuel to flow from a supply source (the fuel rail) into the engine's intake manifold. Injectors are designed so that only a precisely controlled amount of fuel, measured in weight or volume, can flow through in a given amount of time. The injector doesn't actively inject the fuel so much as open and allow the fuel to flow through.

Injectors are rated in either pounds per hour (lbs/hr) or cubic centimeters per minute (cc/min), but one unit can be converted very easily into the other since the weight per volume (specific gravity) of gasoline is fairly constant.

The amount of fuel that can flow through an injector in a given time varies with fuel pressure, so injectors are usually rated at a standard pressure of 3 bar (three times the pressure of the atmosphere, or 43.5 psi). Your stock fuel system does not necessarily run at that pressure (as you'll see), but this allows you to compare injectors without worrying about the pressure at which they are rated. Of course, the fuel pressure regulator increases pressure with boost and lowers it with engine vacuum, ideally keeping the pressure seen by the injector the same at all times.

Fuel injectors are controlled by the engine's ECU, which sends an electrical pulse that opens the injectors. The length of this pulse controls the air/fuel ratio inside the engine's combustion chambers. The longer the injector is held open (for a given fuel pressure), the richer the mixture, while shorter pulses result in a leaner mixture. When idling or under cruise conditions (small throttle opening like on the freeway), the injector pulse is very short and only a small amount of fuel is injected. When the

Fuel injectors are actually measured to determine their flow rates. A good injector supplier, like RC Engineering, will measure every set and make sure that the number scribed on the body accurately reflects a particular injector's flow rate, and that a set of four is closely matched. This makes for a better-running and easier-to-tune engine.

driver gives the engine full throttle for acceleration or when boost comes on, the pulse becomes longer to supply the additional fuel that is needed.

The injectors' flow rate is the primary restriction to fuel output, but it is related to another measurement of relative flow: the duty cycle. Duty cycle is simply the percentage of the available time that an injector is open and squirting fuel. The available time is the window between one injector open pulse and the next. In a four-stroke engine this is two entire revolutions of the engine, since three cycles, compression, power, and exhaust, must occur before the injector can be fired again during the intake stroke for each cylinder. Some tuners talk of duty cycle in terms of the intake valve open period, but really the entire cycle is available for fuel flow, since excess fuel will simply pool behind the valve and be sucked into the chamber when the valve opens and air goes rushing past.

As you can imagine, the time available for the injector to be open decreases as the RPM increases, but the duty cycle must increase if you increase an engine's airflow at high RPM. This is important because as duty cycles increase, injectors can get flaky; they have a small electrical solenoid valve inside, and this valve can overheat and jam open (or closed!) if the injector is kept continuously open. Although fuel injectors are flow-rated at 100-percent duty cycle (full-open), most tuners and injector manufacturers, such as RC Engineering, recommend that you not run injectors at more than 80 percent duty cycle. For most injectors, that is the preferred maximum for long-term reliability.

In some situations, injectors that should be correct for the horsepower desired still cannot keep up. This is due to another characteristic of fuel injectors known as injector offset. The offset is the delay between the ECU sending an open signal to the injector and the injector reaching full flow. If an injector is marginal in size, the engine may require a pulse almost as long as the time available between cycles, leaving no time for the offset.

The offset of an injector also significantly influences how the injector will perform when the engine is idling. An injector with a large offset (slow opening time) will not be able to function at very short pulse widths such as needed when idling. In addition, faster opening injectors are needed for high-RPM engines because the lower offset allows for more of the open pulse to supply fuel to the engine.

These are some of the reasons why Mitsubishi used low-impedance or peak-and-hold injectors in the 4G63t and other high-performance engines. They are easier to control since they can open much faster than high-impedance or "saturated" injectors. The terms "low" and "high impedance" refer to the internal resistance of each injector. Low in this case is less than 3 ohms (2.2 ohms, to be precise), while high impedance measure 12 ohms or so.

You can run high-impedance injectors with a re-tuned Mitsubishi ECU, but it requires changes to the injector delay and sizing. On the other hand, if you're running a non-4G63t ECU, make sure that you don't use low-impedance injectors unless it's designed for them. Some OEM ECUs can burn up if forced to drive the higher-current injectors.

Stock Injectors

The flow of fuel coming out of an injector is less accurate during shorter pulses, so usually manufacturers pick the smallest injectors they can get away with. This is fine until you start increasing the boost or raising the rev limit—the stock injectors are not big enough for much extra power. If you need more fuel than they can supply, the engine will start to run lean and bad things will happen. This is because the flow capacity of the stock injectors has been exceeded. If you are building a new

These are some of the most common stock 4G63t injectors. From left to right, the pink injector is a 560-cc/min unit from an Evo VIII or IX, the black one is a 450-cc/min 2g DSM injector, the yellow is a 510-cc/min from an Evo III or Japanese Galant with a big turbo, and the blue one is 450-cc/min from a 1g manual transmission car.

engine or modifying an existing one, it is always important to determine if your injectors are large enough for the new combination.

Stock USA-market 4G63t engines have come with four different injectors. The 1g manual transmission and all 2g injectors are the same size: 450 cc/min. The 1g automatic cars have 390-cc/min injectors. The Evo VIII and IX have 560-cc/min injectors. All of these ratings are at the same reference pressure of 3 bar, although Mitsubishi used a different rail pressure in each application: 43.5 psi for 1g AT, 36.3 psi for 1g MT and 2g, and 43.5 psi for Evo.

As mentioned before, all 4G63t injectors are peak-and-hold type with a resistance of 2.2 ohms. All are interchangeable, since they are the same length and have similar connectors. Generally they are classified on the basis of the color of the plastic top (different flow rates have different colors).

Bigger Injectors

Now that you know what stock injectors were supplied with your 4G63t, you can think about upgrading. Choosing the right injector size is often difficult for a novice tuner because there are very few good explanations of the relationship of injector size to performance. But don't forget that changing injector size always requires a change in ECU tuning or some other way to compensate. You can't just swap in larger injectors and expect your engine to run right.

The problem is a simple one: Find the smallest injectors that work in your engine combination, meaning that they are large enough for the engine's power demands without being too big. Injectors that are too small will not be able to supply enough fuel at high RPM and boost

levels, while overly large injectors will make a steady idle difficult or impossible to tune, or cause poor low-speed or off-boost performance in a race engine.

The main principle governing fuel injector size is the fixed relationship between the pounds of fuel that an engine consumes and the horsepower that it produces. This relationship is called brake specific fuel consumption, or BSFC, and differs between types of engines. BSFC ranges between 0.4 and 0.7 pounds per horsepower, with carefully assembled and optimized engines at the low end of the spectrum and turbocharged engines at the high end of the spectrum.

The lower the BSFC the more efficient the engine. Turbocharged engines have a higher BSFC because they need more fuel to produce maximum power than NA engines, partly due to the cooling effect of rich mixtures that allow engine tuners to use more advanced timing. BSFC is very hard to calculate accurately; to determine BSFC, the tuner would have to measure the amount of fuel consumed by an engine on the dyno and record the horsepower that it produces at the same time.

An interesting side effect of this is that if we know the BSFC of an engine, we can calculate the horsepower that it will produce. For the purpose of injector sizing, however, we can estimate BSFC based on what similar engines use. For our purposes, most turbocharged engines have a BSFC of between 0.55 and 0.65, depending on a number of factors, including the fuel used. In states where the only fuel available is 91 octane, more fuel is needed to combat knock, which increases BSFC. To ensure that the injectors you need are on the safe side, it would not hurt to use the higher BSFC numbers when calculating your engine's needs.

With this estimated BSFC and a realistic horsepower estimate, we can calculate the injectors needed for a particular combination. The more accurate the horsepower estimation, the more accurate the injector size you'll end up with. The best information comes from careful review of the dyno plots (not just dyno numbers) posted on Internet forums like evolutionm.net, dsmtalk.com, and dsmtuners.com. By extrapolating from wheel horsepower numbers laid down by cars set up similar to yours, you can figure out the best injectors to use for your combination.

The basic formula multiplies the estimated horsepower by the BSFC to get the engine's total fuel demands. The result is divided by the number of cylinders to get the fuel requirements of each cylinder. To get the size of injector needed, the number of cylinders is multiplied by .80, or 80 percent, to account for the injector's maximum duty cycle. The final injector size number is in lb/hr, and can be converted to cc/min by multiplying by 10.515. The complete formula is below.

To illustrate these injector sizing calculations, let's use a 1g motor with stock cams and rev limit. With the stock 14B turbo, injectors, and fuel system, and running a little extra boost (14–15 psi), 200 hp at the wheels (or about 220 hp at the crank) is possible.

Experience shows that at this point, the stock injectors will be maxed out or close to it; let's look at

---

Max hp X BSFC / (# cyls X 0.80) = fuel injector size (lb/hr)

CC/Min = lb/hr X 10.515

> *Max hp X BSFC / (# Cyls X .80) =*
> *220 X 0.6 / (4 X 0.80) =*
> *132/3.2 =*
> *41.25 lbs/hr X 10.52 =*
> *433.95 cc/min*

the numbers to see just how close to the edge this is. The BSFC number we'll use is 0.6 since it is a stock turbocharged engine with all of the inefficiencies and rich tuning that comes with it (using some kind of fuel management to get the mixtures a little leaner will lower BSFC somewhat)—see above.

Yikes! The stock 1g injectors are rated at only 450 cc/min, and flow at a pathetic 411 cc/min at the stock fuel pressure, which is lower than our calculated numbers. This shows that even if you upgrade the fuel pump, the stock 1g injectors are too small for boost increased beyond stock. The 1g injectors should be replaced with at least 550cc/min Evo injectors, and the fuel pressure should be increased to 43.5 psi before you start increasing the boost.

What if we put a bigger turbo on our example engine, say, an Evo III 16G running at 20 psi? We already know from stepping through the calculations that the stock injectors don't have any headroom for power increases. With the new turbo, more flow will certainly be required. But how much more will we need? From looking at dyno graphs of similar engines we can see that the power output will be around 280–300 hp.

> *Max hp X BSFC/(# cyls X 0.80) =*
> *300 X 0.55/(4 X 0.80) =*
> *165/3.2 =*
> *51.56 lbs/hr X 10.52 =*
> *542.44 cc/min*

We can assume that the engine will be tuned carefully to take the most advantage of every bit of fuel, which means it will be leaned out somewhat, and the timing advanced to near the max torque point. This will probably decrease the BSFC, so 0.55 is a more realistic number, giving a little bit more injector headroom. See below for a recalculation of our injector requirements using the higher horsepower estimate.

Recalculating our injector requirements shows us that our new combination needs injectors that flow 52 lb/hr or 543 cc/min (at the fuel pressure you are running) to keep up. The nearest common size of injector is 550 cc/min at 43.5 psi, which will work just fine for this combination. However, knowing how power-hungry enthusiasts get, most people would be better off buying 600 or 650 cc/min injectors and living with a slightly rougher idle. With all of these injectors you should be running an upgraded fuel system at 43.5 psi.

If you find, as in this example, that you need larger injectors to keep up with increased engine power, the first step is to decide on a method to increase injector flow. One way is to swap the injectors with larger units, this is the preferred way in most cases.

For increases up to 5–10 percent, fuel pressure can be increased instead of using larger injectors—all that is required is an adjustable fuel pressure regulator. There are limits to fuel pressure increases though, and if pressure has to be 55 psi or greater it is probably time for larger injectors. Your fuel system is designed for a certain pressure, roughly 43.5 psi, and increasing pressure significantly above that can lead to problems with leaks and fuel pump damage, as well as a drop off in total flow as inefficiencies in the system are exposed.

Many successful cars have been built with a high-flowing fuel pump and an adjustable fuel pressure regulator (FPR)—just raise the pressure to

Aftermarket 600-cc/min injectors should be close enough to our estimated 543-cc/min that they should not be a problem. They might give a slightly rougher idel, but can still be tuned to run well.

An aftermarket adjustable fuel pressure regulator like this quality AEM unit will allow changes in fuel pressure to compensate for greater fuel demands as well as keep fuel pressure more steady in DSM cars, which have a too-small regulator to begin with.

get more injector flow and you're set for moderate power levels, say, up to 250 to 300 hp. This kind of setup is a little harder to tune than larger injectors, and it puts additional strain on the fuel system that isn't optimal, but it's one way to do it.

Since manufacturers give the fuel pressure that an injector is rated at, we can easily calculate the fuel pressure needed to achieve a higher flow rate by using the following formula, and see if the result is too high. The new flow rate is equal to the old flow rate multiplied by the square root of the new pressure divided by the old pressure. This is much easier to see as a formula:

$$New\ Flow = \frac{Old\ Flow\ X}{\sqrt{(New\ Pressure\ /\ Old\ Pressure)}}$$

Or, to find the pressure needed for a given flow rate, use:

$$New\ Pressure = [(New\ Flow\ /\ Old\ Flow)\ X\ \sqrt{Old\ Pressure}]^2$$

We can apply this formula to the situation above—a 16g-powered 1g—to see how it all works together. We just need the desired flow rate (543cc/min) and the current flow rate (411cc/min), as well as the current fuel pressure (36.6 psi)—see below.

By using our formula, we can see that our new pressure has to be nearly 64 psi. It takes a lot of fuel pump capacity to supply enough fuel for 300 hp at 64 psi, which means we probably shouldn't use a fuel pressure regulator set to a higher rail pressure. We will have to buy larger injectors to supply our engine.

$$[(543\ cc/min\ /\ 411\ cc/min)\ X\ \sqrt{(36.6\ psi)}]\ 2 =$$
$$(1.32\ X\ 6.05)2 = (7.986)2 = 63.77\ psi$$

As far as mechanically fitting larger injectors, suppliers such as RC engineering make drop-in DSM injectors with the right connectors and O-ring grooves. The injectors shown here range from 550 cc/min to an astounding 1,200 cc/min, or enough for a 600-hp 4G63t.

Always use new O-rings when you swap injectors or change fuel rails. There is a lot of fuel pressure in the rails, and a leaky O-ring could cause a nasty, and dangerous fuel leak. The O-rings at the bottom end of the injectors should be replaced too: They're a common source of boost and vacuum leaks.

The stock DSM fuel pressure regulator's biggest problem is a lack of flow. If you upgrade the fuel delivery to the rail via a larger pump, fuel pressure can become too high at low fuel demand and mess with low-speed fueling. The general rule is to upgrade the regulator when you upgrade the fuel pump.

Fuel Pressure Regulators and Gauges

Fuel pumps are "dumb"—other than an internal bypass of roughly 70–100 psi, they simply pump out as much pressure as they can. While the pump is the source of the fuel that the injectors see, it does not control pressure; that's the job of the FPR. By bypassing some of the fuel back to the tank, the regulator keeps pressure constant no matter what the injectors' demand for fuel and no matter what the pump's output. This makes the ECU's job easier since it does not have to calculate the pressure in the fuel rail to determine how much fuel is flowing through the injectors.

The Evo fuel pressure regulator is similar to the DSM regulator, but these cars don't have the same problems with rising fuel pressure. Partly this is due to the better regulator, and partly to the improved fuel plumbing back to the tank.

Metering the fuel requires a precise pressure differential between the rail and intake manifold, so the FPR is referenced to intake manifold pressure. As boost comes up, so does fuel pressure. On the other hand, when the manifold pressure is negative, fuel pressure drops in order to keep the pressure across the injector constant. The stock 1g regulator seeks to maintain a differential of 36.3 psi, while the 2g and Evo regulators target at a more standard 43.5 psi.

If you upgrade the fuel pump to a larger unit, the higher flow level will quickly overwhelm the stock regulator at idle and low fuel demand conditions. The regulator won't be able to return enough fuel to keep the pressure down, which will result in high fuel pressure and a rich mixture. The 2g AWD cars have a restrictive return system that makes this problem worse. Of course that pressure will drop as the injectors spend more time open, which makes it hard to tune the engine to keep a constant air/fuel ratio.

You won't have this problem if you're using the stock fuel pump, but

if you've swapped your pump, an easy way to check for an overrun regulator is to check the fuel pressure at idle. Anything appreciably above stock pressure (taking into account manifold vacuum) indicates that there is too much fuel flow for the stock regulator and return line to handle.

The solution is often an aftermarket fuel pressure regulator, unless you're dealing with a 2g AWD car. If the problem is the restrictive fuel return line, even an aftermarket regulator won't help until you change the return plumbing to dump more directly back to the tank. Aftermarket regulators have larger inlet and outlet ports than stock, which means they can bleed off pressure to the return line faster than stock. This keeps pressure constant at all times even with an upgraded fuel system.

Aftermarket fuel pressure regulators are also adjustable. This means you can adjust them to hold any fuel pressure that you want, for an additional tuning variable, as noted above. There are a few good brands of pressure regulator. The best appear to be the Aeromotive and AEM units. Don't waste your time with "rising rate" regulators that increase fuel pressure faster than boost pressure increases, because they are almost impossible to tune. They require compensating tricks in the ECU programming that can make your tuner's job difficult.

Lots of people use small fuel pressure gauges and mount them permanently under the hood, but they aren't really necessary and they can even be dangerous. They have a tendency to leak after a few years and even the manufacturers of these gauges don't recommend mounting them permanently. Setting the fuel pressure is not something that you have to do often, so don't worry too much about not having a permanent gauge mount.

If you buy an aftermarket adjustable regulator, you'll also need to have a fuel pressure gauge to adjust it. It should go without saying that you should never install a mechanical fuel pressure gauge inside the passenger compartment (which would mean you'd have fuel flowing inside the car). The only gauges you should ever use are electrical, with a separate sender under the hood.

The best place to mount a fuel pressure gauge on a car without an aftermarket regulator is at the fuel filter banjo fitting. B&M makes a kit designed for just this purpose, but there are others available. Most aftermarket regulators provide a dedicated port for measuring pressure that you can use.

Incidentally, the stock Evo fuel system doesn't need a regulator if you change the fuel pump and keep the stock ECU. The factory fuel pump drive circuit uses a two-stage voltage system that reduces pump flow at idle, preventing it from overwhelming the FPR. The regulator itself is also a little better, which means that an upgrade isn't necessary on an Evo 4G63t until power levels and fuel flow needs go way beyond pump-gas levels, or if you want another tuning variable. That doesn't stop people from installing them, but regulators probably aren't helping those people much.

The problem with aftermarket fuel pressure regulators is bolting them to the stock fuel rail and fuel lines. The stock DSM fuel rail has a two-bolt flange on the end, so you need to buy or fabricate an adapter for standard pipe thread or AN fittings.

This aftermarket fuel rail has a simple bolt-on adapter fabricated from a weldable AN fitting and the remains of the stock FPR. Stock fuel rails can have an AN adapter tig-welded directly onto the end of the fuel rail, or the regulator end of the rail can be tapped to use an adapter fitting with a little sealant.

Other than the issue with adapting an FPR or aftermarket braided fuel line, the stock fuel rail is plenty large enough inside to handle more power than your average 4G63t is going to be putting out, so keeping them and adapting a regulator is probably the better (and cheaper) choice for most people.

Under 500 hp, an aftermarket fuel rail isn't necessary. But if you've got money burning a hole in your pocket, or just a desire for more under hood decoration, knock yourself out. If you just want to make decent power without trouble, keep the stock rail.

Fuel Lines

The stock fuel lines in all 4G63t-powered vehicles are fiber-reinforced rubber hoses with swaged ends, connected to hard steel lines of approximately 8-mm outside diameter. They aren't sexy, and they aren't high tech, but these lines work very well for moderate horsepower levels. The rubber hoses on the end of the hard lines have a tendency to decompose after a time and most 1g and some 2g cars should definitely have them replaced by now. Since it's a safety issue, replace them at the first opportunity.

Lots of people upgrade the under hood section of the fuel lines to braided stainless hoses with threaded AN fittings on the end. There are a couple of reasons that they do this, the first being the larger inside diameter of the braided hoses. The second is the ease that this makes installing an aftermarket regulator. Looks are also an aspect, as everyone likes the way stainless braided lines dress up your engine compartment. All of these are good reasons to swap from the stock, and by now possibly grubby, rubber lines.

This leaves the stock fuel filter in place, and for many people this works fine, but when you start pushing the stock fuel system to accommodate 400 hp or more, you'll want to replace it with something that flows a little better. Don't be tempted to use small aftermarket filters that have too fine a filtering media—they will restrict flow even more than the stock filter.

Aeromotive and several other companies make nice alloy fuel filters that you can use, but you will have to track down the special 14-mm (threaded diameter) compression fitting that you will need to adapt it to the stock hard lines. Earl's has one for GM power steering lines that will work but is not optimal. The same fittings will be needed for the tank-to-line fittings, too.

If you want to install a braided stainless fuel line in the engine compartment (top), you need an adapter for the fuel filter banjo fitting and an adapter for the fuel rail at the minimum. It makes sense to install an aftermarket pressure regulator and matching AN fittings (bottom) for each connection at the same time.

The stock fuel filter is actually plenty for most pump-gas power levels. It can flow as well as any in-tank fuel pump can, and it does not present much restriction unless it is plugged. Replace it before modifying your fuel system, using only an OEM Mitsubishi filter.

Some people go whole hog and replace the main hard line with a braided hose. The stock steel fuel line is 7-mm inside diameter, and while it is a restriction, it is not the worst restriction in the fuel system, and it is big enough to support hundreds of horsepower. This isn't the best way to do it; flexible hoses are less reliable than hard lines, and they tend to leak at the fittings after a while. They're also hard to route because of the thickness, but that isn't such a problem with any of the Mitsubishis.

The biggest reason to avoid long flexible lines is the leakage issue. On the street, dirt gets in between the stainless braid and the inner rubber hose, and causes it to abrade and leak. For this reason, it's best to keep braided hose usage to a minimum.

If you are set on replacing the main fuel line with something better, use hard lines with flare fittings on the end. These lines are a lot harder to run than flexible hose, but they look much better, and they work much better on the street. If you're

building a race car and don't mind cleaning or replacing the lines on a regular basis, go for braided lines, but for the rest of us, hard lines are a better option.

Fuel Pump Needs

The fuel pump is the first link in the fuel supply chain that ends at the fuel injector. Without a pump supplying enough fuel, the injectors won't be able to get enough into the intake manifold to meet the engine's power demands. The fuel pump, in fact, is one of the weakest links in the DSM fuel-supply system.

To understand why pump upgrades are needed, we'll look at the problems with the stock DSM and Evo fuel pump arrangement: capacity and voltage. Capacity is the most important rating factor for a fuel pump. Figuring out the amount of fuel flow that you'll need isn't very hard. You want enough pump to supply all four injectors running full open, i.e., 100-percent duty cycle, at your highest boost pressure.

Of course, it's not very likely all four injectors will be full open at the same time, but sizing your pump that way gives you a little headroom to cover voltage loss, clogged fuel filters, and the other little things that get in the way between theory and reality. Fuel pumps are rated in liters per hour (lph) instead of cc's per minute, but since there are 1,000 cc of fluid in 1 liter, the conversion is easy.

The fuel pressure that the pump has to supply is nominal pressure plus boost pressure because of the fuel pressure regulator's reference to boost pressure. For example, at idle there might be only 15 psi or less, but at 15 pounds of boost, the pump will have to supply 58 psi of pressure (the nominal rail pressure of 43 psi plus boost pressure of 15 psi).

To calculate fuel pump demands, the whole formula looks like this:

60 X (inj size (cc/min) X # cyls) / 1,000 = Fuel pump size (lph) at needed pressure

As an example, we'll go back to our stock-turbo 1g engine running 15 psi of boost from the injector example above. We'll assume the same stock injectors (450 cc/min), and we'll assume that you've already installed an adjustable FPR set at 43.5 psi because it's a standard fuel pressure. That means that at full boost, your pump will be pushing against 58.5 psi. So:

60 X (450 X 4) / 1,000 = 60 X 1,800 / 1,000 = 108 lph at 58.5 psi

You will need at least 100 lph of fuel flow at 58.5 psi. According to testing performed by RC Engineering for Road/Race Engineering, a new stock 1g DSM pump can crank out about 100 liters per hour (lph) at a fuel rail pressure of 43 psi. It's enough for a theoretical 250 hp, or about as much power as you can get from the stock fuel injectors as our calculation seems to suggest.

Sounds pretty good, right? Well, that's at only 43 psi, which is much lower than the fuel pressure we need to keep the fuel flowing through the injectors properly. As boost raises, the pump has to work harder and harder to keep up, and the flow rate drops off. At that high of pressure, the stock fuel pump flow drops off to 70 lph, or enough for less than 200 hp! Clearly, though the injectors can be pushed to this level, the pump will definitely have to be upgraded if you want to run 15 psi of boost on an otherwise stock 1g engine.

In general, the fuel pump and wiring change should be among the

From left to right this shows a stock 1g fuel pump, a stock 2g fuel pump (the Evo pump looks essentially the same), and an aftermarket Walbro high-volume replacement. The Walbro will drop in place of any of the stock pumps, but be sure to upgrade your fuel pump wiring if you decide to use one.

first upgrades you do to your DSM fuel system. Providing a good, high pressure and high flow source of fuel will give you a foundation for future horsepower increases up to the pump's flow limit and beyond, if you upgrade the wiring too.

The stock Evo pump is better than the DSM pumps, cranking out over 150 lph at 43 psi and almost 120 lph at 58 psi. For the Evo guys, you've already got an upgraded pump in your tank, and the need for an upgrade is pushed up to a much higher power level (300 hp at the wheels or more).

Fuel Pump Wiring

The second problem with the stock fuel pump is its power supply. The stock wiring on DSMs is woefully inadequate. The pump wiring starts at the ECU, with a low-amperage signal that turns on the main fuel pump relay under certain conditions (generally only after the engine starts to turn over or the starter is engaged). Power for the relay comes via the main ECU harness from the battery; it then runs through several connectors back to the fuel tank via an 18-gauge wire. The pump is grounded through another 18-gauge wire.

All of this small wiring and connectors adds up to a lot of voltage loss by the time it gets to the fuel pump mounted in the tank. According to RRE, the average 1g or 2g DSM will produce less than 12 volts at the fuel pump—sometimes much less, even when the engine is running (and charging the battery). Compared to the battery's nominal voltage of between 13.8 and 14.1 volts, this is a drop of more than 2 volts.

The stock pump that can only manage 100 lph with stock wiring jumps all the way up to 130 lph when it's connected directly to the alternator or battery with a thick, low-resistance wire at 14 volts. At 58 psi, it still

Low voltage really hurts fuel pump performance. Fuel pumps are DC electric motors; their speed and therefore flow varies with the voltage that is supplied to the pump. Since the fuel pump is designed for 14 volts, the 12-volt stock power supply results in a hugeloss of flow—on the order of 25 percent.

holds on to 100 lph, which is (just) enough for our raised-boost engine. Of course that's assuming that the pump is in tip-top shape, but it illustrates just how bad the voltage drop is for fuel flow.

The lesson that comes from this is that before doing any fuel system work, you need to re-wire the pump power supply directly to the alternator. When the engine is running the alternator is the source of voltage in the car's electrical system, not the battery. You can use a wire as small as 12 gauge to get a lower voltage drop, but most people who rewire the fuel pump recommend using a 10-gauge wire. That gives you room for a higher current-draw pump without increasing the voltage drop.

To re-wire your fuel pump, you need enough 10-gauge automotive wire to get from the engine compartment to the fuel tank, a 30-amp automotive relay (the Bosch and Hella units are the most reliable), connectors for the relay, and an ATC fuse holder with heavy-duty pigtail wires. Install the fuse holder directly to the back of the alternator with a ring terminal. The other end of the fuse holder can be connected either to the 10-gauge wire or to the relay connector.

On 1g cars, the fuel pump can be triggered in the engine compartment from the fuel pump test wire located on the firewall harness. On 2g cars, your best bet is to put the relay near the original fuel pump connector above the tank. Trigger it with the old small-gauge wire that previously fed the pump. Whatever you do, make sure that the fuse is as close to the alternator as possible, and make sure that all of your connectors are crimped and soldered to their terminals with the best workmanship you can manage. A failure of just one of these wires will strand you on the side of the road.

The stock Lancer Evolution VIII and IX fuel pump wiring is much better. A second relay feeds voltage to the pump directly from the battery under high loads to prevent the fuel pressure regulator from being overrun. There is a small voltage drop to the pump that could be wired around if you're planning on running huge boost levels, but it isn't needed for most street usage.

If you want to rewire your Evo pump, don't forget that the stock ECU controls the fuel pump with voltage. At low demand, like idle, the pump gets less than 10 volts, while it jumps up to 13 volts at high power levels. To increase voltage add the same wire as for a DSM, but in this case, trigger it with the wiring running to the stock hi-lo voltage fuel pump relay. It's wire 39 at the ECU. This wire is switched to ground when the pump goes into overdrive, so trigger your relay on the ground side with this wire.

Don't forget to install a ground wire in your pump setup, too. The ground wire is just as important as the power feed wire, and any resistance in the long stock ground wires will cause an effective voltage drop. Ground to a fresh bolt near the fuel pump, as well as a ground in the engine compartment, to be thorough.

Upgraded Pumps

If you're planning on running more than 200 hp at the wheels on a DSM, or more than 300 hp from an Evo engine, you'll need to upgrade your fuel pump for sure. The most popular fuel pump swap for all these cars by far are the various pumps manufactured by Walbro: the 190 lph, 225 lph, and 225 lph high pressure.

The specification pressure is at 40 psi, but the high-pressure variants have a stronger internal bypass

Really big fuel demands require really big fuel pumps. This Bosch external pump can supply enough fuel for 400–500 hp. It was state of the art in the 1990s, although modern pumps from the aftermarket can easily outflow it.

spring, so they hold on to their flow rate at higher fuel pressures. They are the preferred pumps for turbo applications like the 4G63t, since most of their power is made at high manifold (and fuel) pressure values. All of the Walbro pumps are available in a bolt-in version to replace the stock Mitsubishi pumps.

There is a cheaper alternative for DSM guys: the stock Evo fuel pump. These can be had cheaply from Evo owners upgrading to the bigger Walbro pumps, so if you're looking for normal street power levels of 300 hp or less with your DSM, they're an excellent pump. They tend to be quieter than the Walbro pumps, too.

Because of the relationship between voltage, resistance, and current, installing a larger pump will make any voltage drop in your setup worse, so you should always consider a wiring upgrade to be a necessary part of installing a fuel pump on a DSM. Why go to all the trouble and expense of installing a higher-pressure fuel pump if it's just going to run on 12 volts or less?

Dual Pumps

If you're going for massive horsepower, like above 500 hp at the wheels, one in-tank fuel pump won't

cut it. The only way to increase flow beyond what a single in-tank pump can crank out is with either an external aftermarket pump like the Aeromotive units, or dual in-tank pumps. Since two such pumps will easily overwhelm even most aftermarket fuel pressure regulators, the best approach is a staged system that brings on the second pump only when needed.

Both pumps are installed in the tank on the stock bracket with some modifications. Add a 5/16 hose Y-pipe to connect the outputs of both pumps to the tank outlet. Then run the wiring for the second pump through a fuel-rated bulkhead connector (check out marine equipment supply shops for these).

Two fuel pump relays can handle the current needs of the two pumps, but only one pump (the one that uses the factory pump connector) should be triggered by the original fuel pump wiring. The second pump should be set up with a switch that brings it on-line only in extreme conditions, but before the first pump begins to run out of fuel.

Use a fuel pressure gauge to check for the differential between manifold pressure and fuel pressure—the second pump should come on when that difference starts to drop from its idle level (usually it's above 20 psi of boost or more depending on the setup).

The second pump relay can be switched on with an adjustable pressure switch, called a Hobbs Switch. Adjust the Hobbs Switch to bring it on at your chosen boost point, and your fuel demands have been resolved. Using dual pumps has at least one disadvantage (that it shares with water/methanol injection systems)—if one of the pumps fails, fuel supply will drop off and the engine will run dangerously lean if it is tuned to take advantage of the second pump's extra flow.

ENGINE MANAGEMENT AND TUNING

In a modern EFI engine like the 4G63t, control comes from an engine control unit, or ECU, that incorporates the input from many different sensors to determine airflow and thus how much fuel should be injected to produce a desired air/fuel ratio. It controls the ignition system at the same time, keeping track of revs and airflow to determine timing advance or retard based on load and engine speed.

All stock 4G63t ECUs use similar sensors and engine control outputs, but the internal programming, connectors, and tuning of each model varied quite a bit. ECUs changed from year to year, sometimes significantly, through the DSM production. Mitsubishi also created dozens of ECU part numbers with unique programming for each year, car, and market (California or federal).

This does not mean that each of those ECUs cannot be swapped, though. All of the ECUs are interchangeable, but some, like the 1990, can only be interchanged after a few minor changes. Several of the sensors changed over the years as well. The EVO ECUs are not interchangeable with the DSM ECUs, and are different

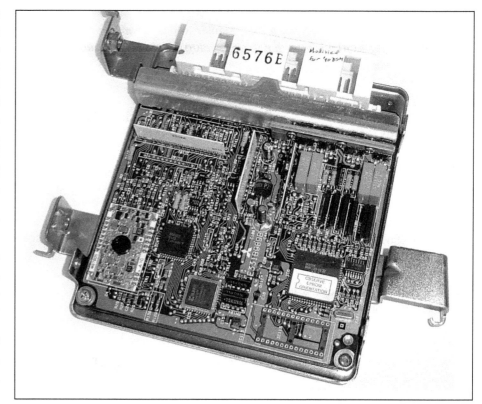

The 1g ECU has an 8-bit processor. The ECU shown here was an EPROM-equipped version, which is the one needed to run DSMLink or an aftermarket chip. It's had a socket installed for easier tuning.

enough that the Evo VIII and IX ECUs cannot be swapped.

1g DSM cars, including the Galant VR-4, all had a similar 8-bit ECU that can be swapped with a few notes. Ignoring tuning differences, FWD and AWD ECUs are interchangeable. The same goes for automatic and manual-equipped cars. The ECUs are interchangeable, but

The various 1g ECUs are similar, but not all of them have an EPROM. You will have to pull the lid off and look in the bottom right corner to find out if yours can be tuned. This one has been "chipped" with a modified chip programmed for a particular combination of parts. Swapping chips is not as convenient as reprogramming a DSMLink adapter from a laptop, but it works just as well.

The 2g ECUs have a different processor, but can be tuned similarly to the 1g. Only 1995 non-OBD-II ECUs like this one have an EPROM that can be removed. The ECU plug is similar to that of an Evo, but you can't just plug one in and expect the engine to work.

Later 2g ECUs are designed with an implementation of OBD-II to follow California smog laws. Your car must be equipped with an OBD-II if you live in states that do smog testing via the OBD-II port. The only way to modify one of these ECUs is with an adapter board like this one.

the auto-trans cars have smaller turbos and injectors.

This means that the manual ECU can be used to add power to a slushbox-equipped DSM with matching performance parts—just make sure that whatever ECU you use matches the engine more than the chassis. ECU tuning vendors like DSMLink, DSMchips, and KeyDriver can all accommodate automatic/manual swaps as well if you have a concern about compatibility.

The 1990 ECU is the least compatible of all 1g ECUs because it uses a different tachometer signal and so requires some creative parts swapping. In addition, two of the pins at the ECU harness were swapped between 1990 and 1991. The wires leading to pin 14 (green, air flow sensor active filter reset) and pin 6 (green/white, idle position switch) have to be reversed to use a 1991–94 ECU in a 1990 car, or the reverse. In addition, the ignition power transistor from a 1991–1994 1g has to be

used to properly drive the tachometer, though no changes have to be made to the ignition when using a 1990 ECU in a later car.

2g DSM ECUs are similar. Automatic/manual and FWD/AWD can be swapped. The 1995–1996 ECUs are interchangeable, although FWDs manufactured after July 1, 1994, use an ECU that has a speed limiter that is not present in the AWD and early 2g FWD ECUs. The 1997–1999 ECUs use a different Camshaft Angle Sensor (CAS) from the 1995–1996 ECUs and require either rewiring to swap coil wires, or adding an electronic signal inversion circuit to utilize a different CAS.

The wiring harness and pin assignments are the same. The inverted CAS signal means that the ECU will fire the plugs and injectors 90 degrees out of phase to the engine stroke cycle. The same problem arises when a 1g engine is used with a 1995–1996 ECU, since the 1g CAS and the 1997–1999 CAS produce similar reference signals.

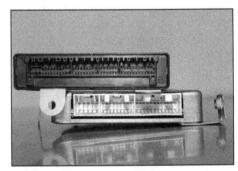

The 1g ECU connector (shown on the bottom) and 2g ECU connector (on top) are not interchangeable. The 2g and Evo connectors are the same but with some pins swapped. Make sure that you have the right ECU for your harness if you plan on swapping them around.

The plug wires are easy to swap, but the injectors are more difficult. Some engines will run just fine with the injectors out of phase, but others will not. The easiest way to fix the problem is to reprogram the ECU, though the injector wires can be swapped at the injector connector. Pin 1 (injector 1) is swapped with pin

Originally required for smog reasons, the OBD-II interface has become the modern tuner's preferred method of reading, datalogging, and modifying the Evo's ECU. With the right cable, the OBD-II port can be used to make the engine run perfectly with aftermarket injectors, MAF, turbo, and other parts.

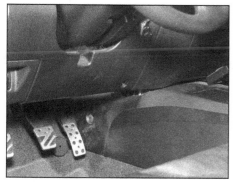

The OBD-II (onboard diagnostics level 2) port on an Evo is located under the right edge of the driver's side lower dash. It's a small black connector with female pins.

15 (injector 4), and pins 2 and 14 (injectors 2 and 3) are also swapped.

Evo ECUs are very different from DSM ECUs. The connectors are similar but the pinouts are different and they incorporate some features, like an immobilizer, that make them hard to swap. The Evo IX ECU and Evo VIII ECU are not directly interchangeable because of the extra circuitry needed for the Evo IX's MIVEC control.

One of the best things about the Evo ECU is its ability to be reprogrammed through the OBD (onboard diagnostics) port. With the correct

cable (available from Tactrix) and software (available from OpenECU and other sources) the Evo ECU can be completely reprogrammed.

This gives the Evo's ECU the ability to control many different engine combinations and boost levels without a problem, and makes tuning much easier than the DSM ECU (with the exception of DSMLink, which we'll get to later).

A few people have managed to swap an Evo VIII ECU into a 2g DSM; the connectors are the same but the pinouts are different enough that the easiest solution is to use a patch cable. In addition, an ECU immobilizer box and programmed key have to be used to use the Evo ECU, unless someone reverse-engineers the immobilizer code in the ECU. It's not a swap for the faint of heart, but it's an interesting possibility.

Stock ECU Sensors

In order to determine how much air is going into the engine, and therefore how much fuel to inject, the ECU monitors several main sensors. The abbreviations for these sensors are industry standard, but the following list will help you keep them straight while reading this chapter:

MAS Mass Airflow Sensor: The Mitsubishi MAS is known as a Karman Vortex sensor, which outputs a variable square-wave frequency depending on the speed, and therefore volume of the air flowing past. The frequency gets higher as the airflow increases. Sometimes called MAF, but MAF is generally used for mass airflow sensors that don't require barometric compensation. Two more sensors, to determine air pressure and temperature (listed below), are needed by the ECU to convert the speed/volume signal into

The MAS is mounted in the intake plumbing. Usually it isn't so visible as this one, buried under the stock air box. The MAS housing also includes the Barometric Pressure Sensor (BPS) and Intake Air Temperature (IAT) so that the ECU can apply correction factors to the MAS output.

a measure of the true mass of air going into the engine.

BPS Barometric Pressure Sensor: The BPS monitors the pressure of the ambient (outside) air. This varies with weather and altitude. This is also used to modify the raw air volume signal coming from the MAS, since air pressure and temperature determine the mass of a given volume of air. The IAT (below) is the other factor used by the ECU to tweak the MAS signal to match actual mass airflow.

IAT Intake Air Temperature Sensor: The IAT tells the ECU the current intake air temperature—this is one of the two inputs used to calculate mass airflow. The air temperature signal is also used to determine when to retard timing on hot days, and how to adjust fuel and timing needs to compensate for a cold engine.

MAP Manifold Absolute Pressure Sensor: Sometimes called a MDP or Manifold Differential Pressure sensor. The MAP/MDP measures pressure inside the intake manifold. Not all ECUs have a MAP sensor, and the 2g's

The MAP is mounted to the intake manifold. Its function is to help the ECU control the EGR valve in DSM cars, though there are some JDM ECUs that use the MAP for boost control.

The CAS is mounted in different locations on the camshafts depending on the year of the motor. This 2g CAS is mounted on the right end of the intake camshaft, but it may be mounted on the left end of the intake came, or on the exhaust cam of an Evo engine.

If you want to install a 1g engine into a 2g car, you can also use a stock 1g CAS with an adapter harness like the one above. The lower adapter harness is for adapting the 1g CAS to the later 2g ECU harness. Don't forget to reprogram the ECU or switch the firing order at the injectors and coils.

MAP sensor is used only for checking the function of the EGR system. Some JDM Evo ECUs use the MAP sensor for engine boost and load calculations.

CAS Camshaft Angle Sensor: The CAS tells the ECU what the current position of the camshaft so that it knows where the engine is in the four-stroke cycle. It runs off of the intake camshaft only, but because

The stock early 2g CAS can be installed on a 1g head if you're doing a 1g head or engine swap into an early 2g car. It requires some moderately precise welding and machining to the end of the head but it's doable by a determined home mechanic. The biggest advantage to doing it this way is to use the early (1995–1997) 2g sensor instead of having to buy the later one or do a wiring swap.

The CTS is mounted on the thermostat housing. It's the larger one of several sensors in this area (the other is for the dash-mounted temperature gauge). Make sure to use the CTS that matches your ECU if you do any engine or head swapping.

the cams are synchronized via the timing belt, it does not matter. The 1g CAS is adjustable while the 2g CAS is not. The 1995–1996 CAS is mounted on the opposite end of the head (under the timing belt) from the 1g and 1997+ CAS. Evo IX MIVEC engines have two CAS sensors to keep track of the variable intake cam's timing.

CTS Coolant Temperature Sensor: The CTS is inserted into the engine's coolant passages. This is

The TPS can be found on the throttle body. It tells the ECU the current throttle position to help with many calculations including acceleration enrichment and idle speed. Be sure to adjust it according to your shop manual if you disturb it or the connectors.

the primary sensor for determining if the engine is cold, at normal temperature or hot.

TPS Throttle Position Sensor: The TPS tells the ECU the current position of the throttle plate in the throttle body. It is used as an input to a table for tip-in enrichment, as well as for coasting and idle.

The CPS can be found on the end of the crankshaft under the timing belt. It simply tells the ECU which pair of cylinders (numbers 1 and 4 or numbers 2 and 3) are at top dead center at any moment. It's used to detect misfires and provide an accurate engine RPM reading.

CPS Crankshaft Position Sensor: A separate, distinct CPS is only present in the 2g and Evo engines. It tells the ECU when to fire the injectors. There are only two CPS pulses per revolution of the engine. The 1g CAS also contains a four-hole disk to provide a CPS signal as well as the CAS signal. A 1g CAS can be used to replace the 2g CAS and CPS in most cases with some minor rewiring or reprogramming the stock ECU. This would be useful for a 1g engine into a 2g swap, or for making the 2g CAS adjustable to add timing. With recent advances in programmable ECUs making the CAS adjustable is not necessary.

Limits of the Stock ECU

Since the ECU controls the addition of fuel to the air coming into the motor, this means that installing any modification that significantly increases airflow and required fuel (larger turbo or more efficient intercooler) will require ECU tuning of some kind.

On the other hand, if you don't plan to modify your DSM engine very much, the stock ECU will work just fine. It has enough flexibility built in to handle things like larger air cleaners, a few PSI more than stock boost, free-flowing exhaust and ported manifolds, but it won't be able to do much beyond that. That's because the stock ECU programming is a compromise designed to enhance emissions, cold start drivability, hot-weather performance, and reliability.

If you're planning to increase boost beyond 15 psi or so on your DSM, or 20 psi on your Evo, or change major parts like the turbo, short block, or cams, the stock ECU programming will not be able to keep up. Anything that radically alters the airflow characteristics of the engine will confuse the stock ECU unless something is done to compensate.

The ECU contains lookup tables referenced to airflow and RPM. If you change the engine's airflow at a given RPM, you will change the area of the lookup table that the ECU is using. This will result in the wrong ignition timing or air/fuel ratio being used for the current boost level. High boost will also run you right into the worst thing about the stock ECU: fuel cut.

Fuel Cut

Fuel cut is the bane of 4G63t tuners everywhere. In short, it's a fail-safe mode programmed into the stock ECU to prevent engine damage in the event of over boosting. It isn't directly based on boost, but rather on airflow. The engine's MAS tells the ECU how much air is going into the engine, and therefore how much load or power it is making.

When this load reaches above a pre-programmed level, the ECU stops injecting fuel into the engine to stop combustion. If you've increased boost on purpose, the ECU doesn't know that, and your first encounter with boost cut will make you regret modifying your car. In most cases, the engine will suddenly stutter and fall flat on its face as the boost limit is reached. In a worst-case situation, it can run lean enough to "hole" a piston as the injectors get turned off and on, but this is rare.

DSM ECUs jump into fuel cut mode way before the Evo ECU. The DSM ECU will hit fuel cut somewhere in the 17-psi range, or somewhere around the 250–260-hp level. The Evo, on the other hand, has a fuel cut programmed to jump in at a much higher level of airflow. Some cars are able to run as much as 20–22 psi, but most of the time fuel cut occurs at the 19–20-psi level at high RPM. Of course since this is based on mass airflow, the actual point of fuel cut will vary significantly depending on the temperature, humidity, and thus density of the air being taken in by the MAS.

Why Tune the ECU?

The most common reason for needing to tune the engine management system is to accommodate a larger turbo and more boost. By cramming more air into the cylinders with a more efficient turbo, the engine will produce more power. That's fine, but the fuel supply needs to keep up with the greater airflow. The simplest route to more fuel is an upgraded fuel pump and injectors, but that brings with it some problems that must be dealt with.

Larger injectors and increased fuel pressure can't be accommodated by stock ECU programming and will require the flexibility of some way of tuning fuel delivery. Simply increasing fuel pressure or swapping to larger injectors alone will not work, since the ECU will not "know" about these larger injectors. The engine will run rich because the ECU will be sending pulses timed for small injectors, but the new, larger injectors will cause more fuel to be injected.

During low-RPM closed-loop operation the ECU will be monitoring the exhaust for excessive unburnt fuel and will be able to adjust the mixture enough to compensate at least partially, but under load or high operation the engine will run excessively rich. Normally, to avoid these problems, the ECU has to be reprogrammed for the larger injectors or an aftermarket engine control system has to be used.

But before the stock ECU programming could be easily altered, tuners developed a number of ways of compensating for these injectors. They fooled the ECU by installing larger injectors and modifying the scale of the MAS signal to compensate. Some of these techniques are useful today, and some of them have been made irrelevant by modern tuning tools, but we've presented them here because they can be used in a pinch, and might even make sense for a budget tuning project.

Hacking the MAS and MAS Swapping

The most popular method of compensating for larger injectors or eliminating the occurrence of fuel cut before the advent of tunable ECUs was to fool the ECU into thinking that less air is going into the engine. The ECU will then inject less fuel during open loop (boosted) operation, thus leaning out the engine. If you have larger injectors this is exactly what you want.

This also works around the airflow-measuring limits of the stock DSM MAS (particularly the 1g MAS). If your engine's airflow demands are too high, you can run into a situation where turbulence causes the MAS reading to drop even as airflow increases, also known as "dropping counts." At this point the MAS cannot measure airflow accurately and the ECU will not be able to inject more fuel. The result is a lean condition and destruction.

The stock Mitsubishi MAS is a Karman Vortex-type—the more air passes through it, the higher frequency it outputs. The ECU has an internal table that tells it what airflow matches a particular frequency. It only samples a portion of the air stream, and that portion is carefully controlled by the design of the MAS. Different 4G63t engines have different MAS, and Mitsubishi calibrated each one for that particular ECU.

There are a couple of ways of reducing the MAS frequency received by the ECU; some require aftermarket electronics and some don't. The simplest method is to "hack" the MAS. This involves decreasing the percentage of total airflow going through the sensor portion of the MAS, which makes it read lower than the actual amount of air getting into the engine.

The way to decrease the sampled airflow is to increase the non-sampled airflow by removing some of the honeycomb inside the MAS. This has the side effect of reducing the restriction presented by the MAS and will tend to decrease spool time somewhat. This really only works for 1g

If you look down the stock MAS you'll see a number of compartments containing a honeycomb. This straightens the airflow through the MAS and makes it more accurate, but it also presents a restriction.

Remove only the honeycomb obstructing the lower oval passage to allow more airflow. Don't touch the honeycomb in the upper passage of a 1g MAS. Really, you should just swap to a newer (and larger) MAS if you start running into MAS inaccuracies.

and 2g engines because the EVO MAS is constructed differently.

But before you go out and start cutting apart your MAS, keep in mind that the honeycomb is there for a purpose, and removing some or all of it will cause your engine to run rough under some conditions if you don't do anything to compensate. Don't do this with stock injectors or without tunable ECU, since it will

With a DSM you can swap the MAS to get an effect similar to a hacked one. Each generation of 4G63t has a larger MAS; the 1g is the smallest, while the 2g (shown here) is larger (and the same size as the 3000GT MAS). The Evo VIII and IX MAS is the largest of the three.

cause the engine to run leaner. Also, you cannot hack the MAS enough to compensate for very large injectors without totally destroying its ability to accurately measure airflow into the engine.

Some of these problems are impossible to tune around, but it is possible to tune a car with a hacked MAS to run well. The MAS has to be hacked carefully and it requires other parts and a lot of tuning time to set up. Incidentally, the 2g MAS and ECU respond better to hacking than the 1g MAS, but both require some tuning to get the most from it.

Piggybacks and Airflow Computers

The simplest and most common method of tuning the ECU is adding an aftermarket computer that goes between the ECU and the engine. Such so-called piggyback fuel computers can be very useful to the budget tuner, but they are mostly obsolete now. For example, in the old days, tuners got around the stock fuel cut by using a device known as a fuel cut defender. An FCD was (and is;

they are still being made) basically a small computer that intercepts the signal from the stock MAS and clamps it at a particular load level determined by experimentation.

Back then, the FCD was the only way to fool the ECU into flowing more fuel to keep up with larger turbos. This worked great for getting rid of the fuel cut, but it introduced another problem. If the MAS signal was prevented from rising past a certain point, which meant that no matter how much additional air was being taken in, the ECU would not be able to compensate for it by injecting more fuel.

Compensating for this added an additional level of complexity, usually involving a rising-rate fuel pressure regulator (that increases fuel pressure faster than the reference pressure), or some other hack.

A much more sophisticated method of tuning for larger injectors (and larger MAS) uses an airflow computer or piggyback ECU like the Apexi SAFC and similar. These little black boxes intercept the MAS signal on the way to the ECU and modify it according to pre-programmed RPM, throttle position, and MAS frequency points. At each point, the piggyback computer scales the MAS frequency up or down to produce the expected behavior from the stock ECU.

The most common way they are used is to scale down the MAS signal about the same amount as the injectors were up-sized. This has the side effect of increasing the fuel cut threshold, since the ECU bases fuel cut on the amount of air going into the engine. If it sees less air, the higher fuel cut is higher and harder to hit.

The only way a piggyback can be properly tuned is using an accurate air/fuel ratio meter and/or datalogging. The goal is to keep air/fuel ratio in the target range under most con-

ditions. Since piggyback computers have only a few steps, they are much cruder than the stock ECU. The resulting mismatch results in a few problems.

The stock ECU uses three-dimensional tables with many rows and columns to determine the engine's fuel and timing needs. The piggyback reduces the MAS signal, which is one of the most important inputs into determining which table and cell to use. This causes the ECU to potentially use too much timing for the current airflow, or use an air/fuel target that was too lean. The greater the difference between the larger injectors and stock injectors, the greater the difference between actual and modified airflow.

In small doses this is good; more timing advance means more power, as long as fuel octane and boost are at levels that keep detonation under control. The problem occurs when timing gets too far advanced. More airflow should mean less timing, but if larger injectors were used and the

The newest piggyback engine management systems are similar to AEM's Fuel/Ignition Controller. They are much better than the old style of piggyback because they intercept the ECU's commands to the injectors and ignition coils at high loads; everything else is controlled by the stock ECU.

piggyback adjusted to compensate, the ECU wants to use more timing.

With too big an adjustment, the engine starts to detonate. Neither of these solutions (FCD or SAFC) is optimal, but before ECU tuning became widespread, they were the only ways that hundreds of tuners had to compensate for larger turbos and larger injectors. Today's tuners have many more options, though some choose to use a piggyback because of price, simplicity, and the timing that it adds.

Installing a Piggyback

Installing any of these piggyback computers and a lot of other electronics requires splicing into your factory ECU harness. If you're uncomfortable with the thought of cutting and soldering important wires under the dash, you can buy a special splice harness that has female and male ECU connectors on either end of a short bundle of wires.

Installing a piggyback is simply a matter of cutting and soldering to the right wires in the splice harness, installed between the ECU and main car harness. When you want to return the car to stock, simply unplug the splice harness and it's all back the way it was before the installation.

MAF Translator

The stock MAS system works very good for most street cars; it measures air accurately enough for most power needs, can be tuned with a piggyback, and is reliable. The problem with Vortex-style MAS as used by Mitsubishi is that at high power levels it becomes inaccurate and restrictive; most other manufacturers switched to a hot-wire style MAF for this reason a long time ago.

Even the latest Evo X and its 4B11 engine use a hot-wire MAF. Unfortu-

nately, you can't just switch from one to the other. A hot-wire MAF sensor outputs a true mass air signal that's in a different format than what the Mitsubishi ECU expects.

Luckily, someone else has done the hard work of designing an interface between a cheap and easily available MAF (from a late-model GM vehicle) and the Mitsubishi-style MAS output. The MAF translator (MAFT) is just that—it converts one signal to another, and zeros out the barometric pressure compensation for the ECU. It allows you to modify the signal at particular points to produce a different airflow-frequency curve, which makes it a good alternative to a piggyback fuel computer.

Advanced versions of the MAFT can even tweak the timing signal from the CAS to compensate for the increased timing that will result from reducing the airflow signal to the ECU, as well as modify the O2 sensor signal to remove fuel compensation inside the ECU.

The MAF translator allows you choose a MAF-based on-engine power potential and then scale the output to match your injectors and turbo. Unfortunately it has all of the same drawbacks as a regular piggyback fuel computer, namely that it introduces some unwanted changes to ignition timing.

Blow-Through MAF

One of the downsides to the MAS-based ECU that all 4G63t's have is that it measures the air before it gets to the turbo, which means that the MAS/MAF restriction hurts spool (remember pressure ratio explained earlier?). It also means that you can't run a blow-off valve (BOV) that vents to atmosphere; since the air is already metered, a vented BOV will cause the

engine to run rich, stumble and possibly die every time you drop the throttle after accelerating.

The solution to both of these problems is to mount the MAF (only if you're running a MAFT) after the turbo but before the intake manifold. Generally it's best to mount it after the intercooler and after the BOV. This ensures that the coolest, least-turbulent air possible enters the MAS, and that the BOV vents air that has not yet been metered.

A blow-through MAF setup can work very well if it is tuned right, but it cannot be used with the stock ECU and MAS—there are just too many variables in plumbing, MAS location, and air temperature for it to work right. Also, the stock MAS will not meter accurately in the pressure and temperature conditions after the turbo, so it first must be replaced with an MAF. You will also need at least a piggyback fuel computer to tune fuel, and preferably one that allows you to modify timing to compensate for any accidental timing changes that crop up.

Advanced ECU Fooling (Speed Density)

If swapping to a MAF system or moving the MAF isn't enough of an upgrade for you, there's always the possibility of ditching the MAS/MAF altogether. The MAF Translator Pro as well as a few other piggyback computers have the ability to replace the MAS with a MAP (manifold absolute pressure) sensor.

This removes the restriction of the MAS from the inlet tract for better power, and helps high-boost and power consistency. In fact, this is often the best way to control fuel in a turbocharged engine because of the loss of restriction in the intake tract, and because a turbocharged engine's fuel needs are more closely based on manifold pressure than any other factor.

Unfortunately, you can't just replace the MAS with a MAP sensor. A MAP can only be used to find the airflow when the engine's VE is known. Since VE varies by engine speed and manifold air temperature, both of those variables have to be taken into account to calculate airflow.

The piggyback computer does this automatically through the added MAP and temperature sensors and sends an airflow frequency signal to the ECU. Set up properly, such speed-density systems can run just like stock but with the ability to deal with high-power turbo engines.

Reprogramming the Stock ECU

To say that piggyback computers, MAF translators, and speed-density conversions are no good is too extreme. Many tuners had and continue to have success using these methods. It's been proven time and time again that a determined tuner with a SAFC and a lot of time can build a combination that will make power and stay together. If you don't have a lot of money and you have one already, don't automatically toss it out.

The problem is they aren't very sophisticated and they can lead to other problems. There is an easy solution—reprogramming the factory engine management system to be happy with the MAS, injectors, and turbo that you've chosen.

A couple of companies have cracked the stock ECU code, allowing for significant tuneability that did not exist before. Thanks to the hard work of programmers around the world, tuners now have access to all of the original code inside your ECU and the tables that tell it how much fuel and

timing to use for a given combination of airflow and engine speed.

The software in the ECU can actually be recalibrated to tune for larger injectors and open the injectors for a shorter time than stock for the same airflow. If this is done properly, the car will start, idle, and run just like stock with injectors as large as you care to run.

The easiest way to achieve this kind of driving perfection with your ECU is to have someone burn a custom or off-the-shelf "chip" or "map" for your ECU (if you have a 1g or 2g DSM with an EPROM ECU), or have them "flash" a map to your ECU if you have an Evo. They can change these tables to compensate for larger turbos, bigger injectors, and ultimately more power.

The best tuning companies have you fill out a sheet detailing your car, engine, turbo, gearing, and other information that will help them design a chip that matches most closely with the combination of parts under your hood. Most of these maps will still have to be tweaked to run better with your parts, and this is where a MAF translator, SAFC, or speed-density computer really shines.

An inexpensive off-the-shelf map combined with a wideband oxygen sensor and an SAFC for final tweaking can make for a good combination for a high-powered street/strip car.

Do-It-Yourself ECU Tuning

If you don't want to pay someone else to tune your engine, there are a couple of ways to get into the ECU code and modify it to suit your needs. The best method for you depends on your car, ECU type and year, and electronics skills. Tuning the EVO ECU is the easiest because the stock ECU contains the program code and tuning data in flash RAM.

This is similar to the flash cards used in digital cameras and the like, as it can be rewritten indefinitely. The diagnostic port under the steering column gives outside access to the ECU's flash memory, which means that only a few things are needed to access the entire ECU memory space.

To get started, download a copy of the program ECUflash from the Internet and buy a cable from Tactrix. With the Tactrix cable and ECUflash, you will be able to download the stock ECU code to a laptop computer, modify it to your heart's content, and reflash it back to the car. Many people leave the cable in the car so that it can be used whenever necessary to datalog and tweak the ECU code. ECUflash gives you access to almost every aspect of the ECU's performance: rev limits, injector size, ignition timing, and more can be changed with a few keystrokes.

DSM ECU tuning is a little more difficult, and can be impossible, depending on the year of the ECU. Some DSM ECUs have the ECU code burned onto an EPROM chip inside the ECU case. This is a kind

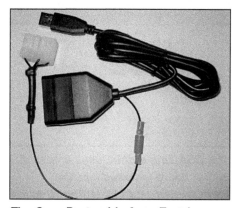

The OpenPort cable from Tactrix allows anyone access to reflashing the Evo ECU. This cable with the appropriate software gives home tuners the ability to tune almost every ECU variable. It's been a revolution in tuning for the Evo.

DSMLink

If you're building a DSM, you won't get past any discussion of ECU tuning without talking about DSMLink. Basically a system to reprogram the 1g and 2g ECUs on the fly, DSMLink has quickly become the most popular and versatile DSM tuning solution. To tune an ECU with DSMLink is easier than tuning most aftermarket standalone ECUs, and installation is much easier since all of the stock wiring, sensors, and actuators can stay in place.

All tuning is done through the onboard diagnostic port, so it's stealthy, too. The integrated design also means that it plays well with piggyback fuel computers, but outside of a MAF translator if you want to run a hot-wire MAF, you won't need one. With a DSMLink, laptop, wideband, and a way to datalog boost, you're in business.

To find out more about tuning the stock ECU, we talked to DSM Link designer, Thomas Dorris.

What are the limits of the stock ECU?

"Generally speaking, the power potential of a particular setup isn't really limited much by the engine management system in place. There are fine adjustments here and there that you make with more complicated systems, but the bulk of your power comes from boost and as long as the ECU is able to achieve your target air/fuel ratio and ignition timing reasonably, there's not a lot of power being left on the table. Modified with DSMLink, the stock ECU has been run well into the 8s in the 1/4 mile at 160 mph. I'm not sure what power potential that suggests, but it's pretty high."

What's the advantage of keeping the stock ECU instead of a plug and play standalone (AEM, etc.)?

"For 2gs, one big advantage is that you retain stock OBDII communications. Another advantage is simply cost. Once you purchase an AEM along with all the supporting sensors and wires, you end up near $2K pretty quickly.

And then you have to tune the thing through countless tables that all interact in different ways, making the task daunting at the very least and completely unmanageable by most. Keeping the stock ECU configured properly to match your MAF and injector size gives the user instant drivability without having to adjust a single thing. They can focus more on performance tuning than just getting the thing to idle because the factory ECU already handles trivial tasks like idle and cruise pretty well."

Would you suggest that beginners try to tune their car themselves? What advice would you give them?

"They have to start somewhere. So yeah, I'd suggest they tune the car themselves. There's really not much they can do with DSMLink that would damage the car. They might make it run poorly and then they'll go undo what they did, but it would be hard to actually damage it. I'd suggest they read the tuning guides, get an understanding of the basic principles, and go from there."

The DSMLink system uses a special circuit board installed into the EPROM-type DSM ECU. The circuit board replaces the original ECU's memory and allows a special software program installed on a host laptop to change the ECU's programming. It gives the tuner nearly as much control as a standalone ECU.

of memory that cannot be rewritten, so the chip has to be de-soldered and replaced with a socket.

All Galant VR-4s and 1990 1g DSMs have EPROM ECUs standard. Among the other DSMs, 1991–1995 1g and 2g ECUs marked with "E" on the case have EPROMs, though unfortunately not all cars have them. None of the 1996–1999 2g ECUs came with an EPROM, though they can be swapped for a tuned 1995 ECU with some wiring changes. Determined tuners have even swapped in Evo VIII ECUs for their flash-tuning capabilities.

Then it can be replaced with either a new chip, a DSMLink module (see sidebar) or an emulator (basically an EPROM that can be programmed on the fly with an external USB cable). To change some parameter of the stock code, just burn a new chip or reload the emulator.

The files needed to decipher the stock DSM code, or "bin" can all be found on the Internet. Mark Mansur's excellent program Tuner-Pro is inexpensive and has definition files for several generations of DSM code. The DSMLink comes with its own software for changing the code,

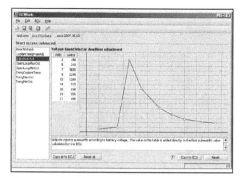

One of the first settings you will have to determine for tuning your ECU is injector size and offset (dead time/latency). Tweak these parameters until the air/fuel ratio correction is 0, and target air/fuel ratios are the same as recorded. Don't be tempted to tweak the MAS scaling at first; get the mixture as close as possible using just the injector scaling.

so it is a more complete solution compared to putting together your own emulator setup.

Standalone Engine Management

DSMLink and other DIY ECU tuning tools have become very good in the last few years, but there are a couple of areas where they come up short. For one thing, the stock ECU does not control boost in a closed-loop manner; rather it depends on the boost control system to take care of the details.

Also, for really high-powered engines, tuners like to control boost, water or alcohol injection, launch RPM, and other factors through the same "box." The stock DSM and Evo ECUs cannot control multiple injectors per cylinder or other exotic fuel systems. It also does not have logic for traction control.

For another thing, tuning stock ECUs is mostly a street car thing. Professional tuners are more common now than they used to be, but many

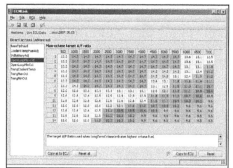

Once the injector and MAS scaling is almost perfect, you can change the target AFR tables in the ECU to get AFR where you want it. The stock ECU programming (as this shows) is ridiculously rich. The high-load cells (bottom right) should be in the 11.5:1 range. Don't change it all at once; creep up to your target AFR and adjust the timing so it doesn't ping.

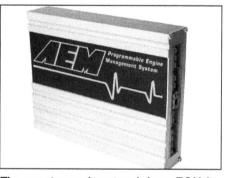

The most popular standalone ECU for Mitsubishis is the AEM system. It's a very good ECU with lots of capabilities, including the ability to self-tune its own VE tables (often the trickiest part of setting up an aftermarket ECU). AEM offers various plug and play variants that use the factory ECU plug and harness for easy installation.

race car shops prefer to stick with one or two aftermarket systems rather than many different modified OEM engine management systems.

For better or worse, many of these shops and tuners got their start before widespread tuning of factory ECUs was possible, the only way to

modify a fuel-injected engine was to swap the entire engine management system for an aftermarket one. After putting in the time to learn the quirks of the aftermarket system, they don't want to take the time to learn different systems.

Aftermarket ECUs (also called standalone engine management systems) generally control the engine with a speed-density strategy. The thinking is, "Why bother with a MAS or MAF when you're changing the rest of the sensors anyway?"

A plug-and-play ECU requires that your wiring harness isn't in particularly bad shape, which is not a safe assumption if you're working on a 1g or early 2g. Building your own harness is easier than it sounds, and is a good option if you want to ensure ultimate reliability for a very high-powered project.

Building Your Own ECU

If you don't have $1,000 or more for a standalone ECU, there is at least one alternative: build one. The darling of the Internet engine tuner, the Megasquirt EFI project, can do almost everything that a bare-bones aftermarket ECU can do, but it requires a lot more work from you.

Far from a plug-and-play setup, Megasquirt comes in many different configurations, from a box of parts to a completely assembled unit. It is a flexible and powerful engine management system but requires some electronics and wiring knowledge to install and use on a DSM.

Thanks to the work of some Internet DSM pioneers, the information needed to configure Megasquirt for use on early 4G63t powered cars is widely available. Configuring Megasquirt to work with the DSM crank and cam sensors and the stock ignition system requires a bit of soldering and adding some jumpers to the Megasquirt circuit board.

Once this is done, it can be used in place of the DSM ECU with the addition of an IAT. To install the ECU, either use a connector cut from a junked factory ECU and build a jumper harness, or replace the harness connector with one that matches Megasquirt's typical DB37 connector.

With proper tuning, Megasquirt can make as much power as any other aftermarket engine management system, so there is no need to worry about power capability. It is also reliable, as demonstrated by the thousands of cars and bikes that have been converted since the introduction of Megasquirt in 2001.

What you won't get with Megasquirt is paid tech support. You'll have to rely on your own tuning and diagnostic abilities much more than with an aftermarket setup, and you will have to figure out some of the installation and tuning details on your own, too. But if you have the time and abilities to use it, Megasquirt deserves more than a passing look as you decide on your engine management plans.

Choosing an Engine Management System

Like most other choices you make when building a performance car, choosing an engine management system is all about compromise. Any engine management system can be made to produce the same power level. It is the engine's other systems, like turbo and fuel system, that are far more important in power production potential than the engine management.

The ECU simply allows you, the tuner, to unlock the potential of those other parts. Because of this, the ECU is best chosen on the basis of

ease of use and cost, with an eye to your power goals down the road. The most important thing that an engine management system can bring to the table is tuneability. The most powerful system in the world won't help you if you cannot tune it. As power levels increase though, the importance of your engine management choice becomes more critical, since the demands on tuning become more precise as well.

If you're looking for a simple, daily driven street car with a few extra horsepower (meaning a 16G or smaller turbo), and don't want to spend a lot of time tuning, you can't go wrong with a simple ECU chip. If you plan to tune the car yourself or have it tuned for bigger injectors and turbo, get a MAFTPro, SAFC, or similar piggyback. Some of the most sophisticated piggybacks, like the Greddy Emanage Ultimate, offer control that approaches the level of standalones or modified OEM ECUs. Ask your tuner or dyno operator for suggestions, since they have probably seen dozens of systems and have their preferences.

For more sophisticated street/strip or track cars, and people who like to tinker, DSMLink is hard to beat for its combination of features and flexibility. Its greatest detraction is dependence on the stock Karman-Vortex MAS, but even that can be overcome with an inexpensive piggyback (the MAFT).

For all-out race cars, there is no substitute for a standalone engine management system. You should expect to build an entire wiring harness for the engine to ensure ultimate reliability, and plan to spend a lot of time tuning. At this level of development, it's possible that you will not be doing your own tuning, so check with your tuner or engine builder first before investing money in a system that he or she doesn't like to tune.

Finally, if you're interested in tuning and tweaking your own ECU and you have a good basic knowledge of electronics or are willing to learn, go with Megasquirt. It has a combination of user-friendliness, features, and cost that is not matched anywhere else. Of course it will require a bit more work than many other solutions, but the end result may be worth it.

ECU Tuning and Emissions

No discussion of ECU tuning would be complete without talking about the issue of emissions testing. What is legal or not is largely a matter of where you live. Some states have strict "visual" laws that require the smog technician to inspect the car very closely and verify that all stock smog equipment, including the ECU and injectors, is present and not tampered with.

Not every modification can be detected by the visual method—many aftermarket injectors, MAF sensors, etc., look just like their stock counterparts. Piggyback ECUs may be detected by the visual method, but a careful installer can find a way to make the wiring look stock, and hide the box somewhere on the car.

The other methods of smog testing include tailpipe emissions tests and OBD tests. Tailpipe emissions tests are a function of ECU tuning and installed smog equipment. A car with a carefully tuned stock or aftermarket ECU should have no problem passing a tailpipe emissions test. The tests normally only concern the low-rpm and low-power portions of the ECU mapping.

As long as the ECU remains in closed-loop stoichiometric fuel control, and everything is working well, the car should pass no problem. Even a standalone aftermarket ECU can be programmed to drive and pass tailpipe emissions test just like a stock ECU, especially if it's paired up with a new, working catalytic converter—preferably stock since many aftermarket cats are not efficient enough to pass tailpipe emissions tests.

The final test performed in many states is OBD testing. The emissions testing station plugs a special computer into the car and queries the ECU for information on its current operating status as well as historical trouble codes and "readiness" codes that detail the status of the emissions control equipment on the car. OBD testing is only mandated for cars made after 1996, which includes 2g DSMs and all EVOs.

To date there are no aftermarket ECUs that can properly respond to OBD queries, so the only smog-transparent ECU modifications that can be done to these cars is reprogramming or re-flashing the stock ECU. For most people, this is only a small handicap, since standalone ECUs are not often used on street cars that must pass emissions tests.

Ignition System

All 4G63t ignition systems consist of a coil driver (sometimes also called, power transistor) that energizes each ignition coil. There are two coils, since all 4G63t engines come with some variant of a "wasted spark" ignition system. They get this name because each ignition coil feeds a pair of spark plugs in opposing cylinders that reach TDC together.

One plug is fired on the compression stroke of a cylinder full of air/fuel mixture and ready to explode and push the piston back down the cylinder. The opposing piston, when the other plug is fired at the same moment, is at top dead center in a cylinder at the overlap between exhaust and intake strokes. Since

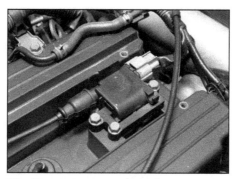

The Evo coils are connected directly to the plugs of cylinders number-1 and 3, while wires fire the plugs in cylinders number-2 and 4. It's a good ignition system, but can be improved on for really high power outputs.

there is no mixture in this cylinder, nothing happens when the plug fires.

In contrast with the fuel and engine control system, most Mitsubishi ignition systems are more than powerful enough for racing and high-performance use. The 1g and 2g DSM ignition systems, in particular, are very good, but the Evo ignition systems are powerful, too.

Any stock 4G63t ignition system with the right plugs and wires will be strong enough for 300–350 wheel horsepower. Above that level, the ignition system might need to be upgraded, but generally you should for symptoms of poor ignition performance (stumbling, misfires, or sudden rich spikes) before changing anything.

Spark Plugs

For such a simple function (to light-off the mixture in the combustion chamber), spark plugs are actually pretty complicated. The electrodes can be made from different materials and can be in different shapes and configurations depending on the application. Most importantly, plugs are designed to draw heat away from the tip at a particular rate. So-called colder plugs have a shorter

path for heat to travel from the tip to the housing. This can help reduce the tendency of an engine to knock. Since hot spots in the chamber can encourage detonation, it would seem that the colder the plug the better from a performance perspective.

Unfortunately it's not that simple—the center electrode needs to reach a certain temperature to burn off fuel and oil deposits that will eventually cause it to foul. Cold plugs may never reach that temperature if the engine is not run at high speed or high load long enough.

On the street, hotter plugs tend to last a lot longer than colder plugs, which is another consideration for manufacturers because of the need for longer maintenance intervals. Racing engines can get away with colder plugs because they are changed more often, and because the engines are run at higher loads.

Electrode material is most important from a wear standpoint. Copper electrodes conduct electricity very well, which makes them almost ideal for spark plugs. Other materials including platinum and

Copper is an excellent electrode material for spark plugs. Unfortunately it wears quickly as electrons move from the center electrode to the ground electrode, meaning that copper plugs have to be replaced as often as every 15,000 miles in a high-performance engine.

The newest plugs have iridium center electrodes with thin platinum coatings on the ground electrodes. This makes them last 100,000 miles or more, but it also makes them very expensive and no better for performance.

This clearly shows why Evo IX plugs (above) can only be used in an Evo IX. Not only is the center electrode made of long-lasting iridium, but the plugs are also longer to get the spark deeper into the combustion chamber. Evo IX plugs will stick out and could hit the piston if you try to use them in an earlier 4G63t head.

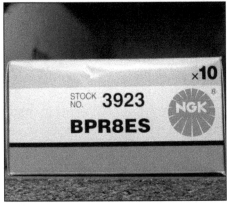

BPR8ES plugs are the coldest plugs that should be used on a tuned 4G63t that is expected to run well on the street. Don't just swap in colder plugs because you've modified the engine since you might end up with more running problems than you solve.

iridium are commonly used on spark plug electrodes.

The stock plug in 4G63t engines varied somewhat over the years, but all were sourced from the Japanese manufacturer NGK. The DSM engines were supplied with copper-electrode plugs. The 1g engines came with BPR6ES plugs, a rather ordinary copper plug with an internal resistor. 2g DSMs came with a dual-ground-electrode variant, the BPR6EKN. For the USA-market Evo VIII and IX, Mitsubishi supplied them with the latest platinum-iridium plugs (IGR7A-G for the VIII and ILFR7H for IX).

In general, for stock or mildly modified engines, stick with the Mitsubishi-recommended plug. There isn't much to gain by changing the type or heat range of the stock plug. The 2g plug is easily replaced by the 1g plug with no loss in performance. In general, copper plugs are best for turbo applications because they provide the best spark, and the electrodes are soft. By contrast, the fragile iridium and platinum electrodes

have been known to break and fall into the combustion chamber because of detonation. In addition, platinum plugs can cause high-boost misfires and stumbling. Copper plugs will not suffer from either problem.

A standard copper plug, BPR7ES, can replace the Evo plug in most cases, also with no change in performance. The copper plug will have to be replaced sooner, but each copper plug costs only 1/10 of the original Iridium plug. That means you can replace 10 sets of copper plugs before you've reached the cost of one set of Iridium plugs.

As for heat range, for most applications, the stock plugs are more than adequate. The "6," "7," or "8" in the spark plug part number refers to the heat range of the plug. The higher the number, the colder the plug. If you plan to push the limits of the fuel you are running, you may consider a switch to a colder plug. However, colder plugs can cause low-RPM misfires and poor off-boost throttle response that you may or may not be expecting.

Only engines running big turbos (i.e., larger than 16G) and 18+ psi might benefit from colder plugs. Heavily tuned DSMs can use a BPR7ES, and high-horsepower Evos can consider a move to BPR8ES plugs (copper) or BKR8EIX (iridium). Stick with the stock heat range and copper plugs unless you have a good reason to suspect that you should be able to run more timing for a given boost level—if your EGTs don't go down after different plugs you probably aren't getting any benefit from them. Hardcore track machines might even need two steps colder than stock, but the same rules apply.

Always check the gap of your plugs before installing them. NGK recommends not changing the gap on iridium plugs, although it can be done if you are very careful not to pry on the conductive part of the electrodes. Most stock DSM ignition systems are happy with a gap of .028 inch. Like everything else ignition related, only change the gap if you have an ignition problem. Very high boost levels (above 20 psi) usually require a smaller gap: as little as .020 to .025 inch to prevent misfires. That is a good starting gap for Evo engines using copper plugs, because the Evo ignition system is weaker than the DSM system.

In general, a larger plug gap produces more complete combustion and can result in more power, all else being equal, but only if the ignition system can fire across the greater gap, and only if the smaller gap is not optimal for the fuel, boost level, and RPM. This is not the case for most DSM ignition systems; the .028-inch gap is all the engine needs to produce optimal power. The voltage from an upgraded capacitive discharge ignition (CDI) system can run a larger gap, but may not produce any additional power. Evos can use more ignition power to fire a greater gap, particularly at high boost levels. Swap over to DSM coils if you plan to increase boost and power much beyond stock.

Plug Wires

In contrast to spark plugs, plug wires are fairly simple parts. DSM engines have four plug wires running from two remote coils located on the intake manifold. Evo ignition systems have only two plug wires; the coils fire the plugs in cylinders number-2 and number-4 directly, with wires running from number-4 to number-1 and from number-2 to number-3. In both cases, good quality wires are a must for performance use. For most street cars, new stock Mitsubishi wires are the way to go. They are not cheap, but they are very good quality and last a long time.

As an upgrade from the stock wires, a good aftermarket set like Magnacore or ACCEL works well. There are other good brands out there, but don't fall for marketing hype and pay too much for some-

Evo ignition coils have mini plug wires (called "boots" by Mitsubishi) that should be inspected every time the coils are pulled. They do go bad, and the symptoms include high-RPM misfires.

thing that's no better than a set of parts-store discount wires. If you have an Evo, don't forget to check or replace the ignition coil boots located under the two ignition coils (a bad boot exhibits the same symptoms as a bad plug wire).

Ignition Modifications

The first modification should be a complete inspection—a bad power transistor or coil will cause intermittent misfires and other problems that can be hard to diagnose and impossible to tune around. If you're fighting potential ignition problems, keep a known good coil and power transistor on hand and swap them for one of the suspect parts. If the problem goes away, the coil and/or transistor is the problem. If it doesn't go away, you have a fueling or tuning problem.

If you are still having ignition trouble that isn't traceable to a defective part, you can consider upgrading. If you have a DSM under 500 hp or slower than an 8-second 1/4 mile, skip right over coil upgrades. The stock

DSM coils are very good. The stock coils can light off mixtures even in very high cylinder pressure environments of lots of boost and high RPM. They can even be used as an upgrade for Evo engines running really big turbos.

DSM coils are very good—the power transistors are strong and the coils produce a hot spark that doesn't get blown out easily by high boost levels.

About the best thing you can do for a DSM ignition system is add an aftermarket ignition amplifier or CDI unit. A CDI is an electronic box that increases voltage to the coil, which in turn increases voltage to the plug for a hotter spark that's more resistant to blowing out.

If you have an Evo, consider swapping to DSM coils and power transistors. Despite the fact that they are older technology, they are stronger than Evo coils and have been proven to be good for more than 500 hp. You will have to fabricate a custom bracket to mount the coils, but the results may be worth it.

Ignition System Swapping

There are some issues that prevent free swapping of ignition systems between different 4G63t engines, but most swaps can be (and have been) done. The 1990 power transistors output a different

Evo coils can be replaced by late-model Montero V-6 coils for a quick-n-easy upgrade. They are almost as good as DSM coils, and plug right in. They're too tall to squeeze under the stock coil cover, but they work very well on high-powered engines.

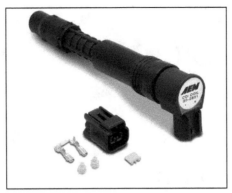

Coil-on plug conversions like this one from AEM are a good idea on paper because they eliminate the plug wires and unreliable connectors of the stock setup. In practice, the stock setup is good to very high power levels.

tachometer signal from other cars, so any swap involving a 1990 car, ECU, or power transistor will require slight rewiring. Take a good look at the wiring diagram of the donor car and compare it to the 1990 parts before swapping them over.

To use a DSM ignition coil and power transistor in an Evo, simply connect the two signal wires to the corresponding DSM power transistor inputs, and tie the two positive and negative power wires together. You may have to adjust the dwell settings in your ECU to make the coils work optimally, but start with the stock DSM settings and tune from there.

Coil-on-Plug Conversions

Over the past few years, many companies have come out with so-called coil-on-plug (COP) conversion kits for 4G63t engines. The idea is a simple one: Ditch the stock wasted-spark coils and plug wires, and replace them with modern "stick" ignition coils that attach directly to each spark plug. It's not a real COP setup because the ECU still fires two cylinders at one time, but each plug gets a positive spark instead of one

getting a negative spark, and the variable of plug wires is eliminated.

In practice, COP is not a magic bullet for high powered engines. The stock power transistors have no problem driving two coils each instead of one, but it's not an ideal situation. The current draw of two smaller coils is greater than each stock coil. In addition, the smaller coils need more dwell or charge time from the power transistors. This is very difficult to do properly with the stock ECU, though

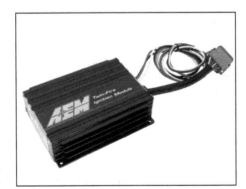

A COP conversion requires the use of an aftermarket 2-channel CDI box like the AEM one. You can get the stock power transistors to work but they can be damaged by the high current demands of the aftermarket coils, and the dwell won't be optimal.

aftermarket ECUs have no problem increasing dwell. So a simple swap from wasted-spark to COP can actually result in less performance than the stock two-coil system.

The way to get around this problem is with a two-channel CDI box, like the AEM or M&W units. There are others on the market, but they all work by firing the ignition coils with voltage charged up in a capacitor instead of depending on the coil to charge in the dwell time available. These boxes don't do as much for wasted-spark ignition systems (see previous) but they are essential to get the most from a COP setup.

Tuning Basics

Tuning is one of the most important steps in building an engine. You can learn to tune yourself if you have plenty of time on your hands and access to more experienced tuners on the Internet, for example, but there really is no substitute for dyno tuning. An hour or two of dyno time and the advice of a good engine tuner are definitely cheaper than broken parts and easier than fighting with intermittent drivability issues.

A dyno pull is always good for showing the weaknesses in a setup, whether it is boost creep or fuel delivery problems that only show up under high load conditions that are hard (or illegal) to maintain on the street. Dyno time is cheap enough now that there is really no excuse for doing without.

One good way to approach tuning is to divide it into day-to-day drivability and full-throttle tuning. Tune the day-to-day, low-boost stuff yourself, and trust the high-boost, high-power tuning to an expert. Some tuners also do an "e-tune" service. This requires you to datalog certain parameters (see below) and e-mail the datalogs to the tuner, who will tweak an off-the-shelf map for you.

This can work well if your tuner is very experienced and your setup is fairly common. If you live a long way from a four-wheel dyno it can be the only choice, and it is probably better than a straight off-the-shelf tune for most people. Just be ready to watch for danger signs when you run any tune that has not been developed by a tuner on a dyno.

Tuning Tools

The most important tool in the tuner's arsenal is careful measurement of all aspects of the engine's performance. If you don't know exactly what your engine is doing at a given moment, you have no chance at changing any of the variables to make it run better. This will require gauges, like an air/fuel ratio gauge and EGT, as well as some way of recording what the ECU is seeing at that particular moment. This is called datalogging. Additionally, some way of monitoring performance will be needed. The simplest way is a stopwatch and an abandoned road, but the ideal (and safest) place to tune is on a dyno.

The most important tool is datalogging. DSMLink and other aftermarket ECUs have datalogging software, or you can use a standalone datalogger like Pocketlogger or EVOscan (for 2g and Evos). All of these software packages work similarly, allowing you to record current RPM, engine load, airflow, temperature, and ignition timing.

Most importantly, they allow you to record air/fuel correction and the ECU's response to knock, two of the most important data points for tuning. The first (correction) is used to monitor the difference between the airflow reported by the MAS voltage and scaling and actual air/fuel ratio,

among other things. This is used to tune injector sizing and latency, as well as MAS scaling. Knock count is a rough measure of the number of knock incidents that the ECU is recording, with each "count" representing approximately 1/3 of a degree of ignition timing removed.

Buy it now or buy it later, but there's no way around using a reliable modern wideband O_2 sensor (like the Innovate, AEM, or Zeitronix units) to check the air/fuel ratio at all combinations of load and RPM. If you want to tune your engine properly, and you plan to up the boost and fueling later, you will need a wideband O_2 sensor. Even if you plan to take your car to a tuner to be set up, a wideband will help you keep an eye on combustion efficiency. A wideband O_2 setup will show you the limits of the stock fuel injectors and fuel system when it goes lean. It will also help you save your engine from damage if you do run out of fuel flow.

Of course if you're planning to pay someone else to tune your engine you won't need your own wideband O_2 sensor, but it could still be useful for keeping an eye on your engine at the track or other stressful situations. An EGT gauge, on the other hand, is a great tool for permanently monitoring dangerous changes in your tune. Once your car is set up correctly, EGT won't vary much under boost, so a spike of high EGT is a good indication that something has gone wrong.

EGT can give the tuner a lot of information about the combustion process. Unfortunately just bolting an EGT thermocouple (temperature sender) into your exhaust manifold without tuning and recording the results will not tell you very much. Generally, EGT ranges from 1,200 degrees F to 1,800 degrees F, with optimal temperatures below the higher

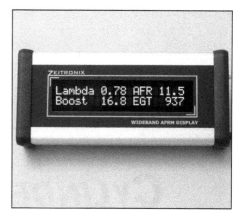

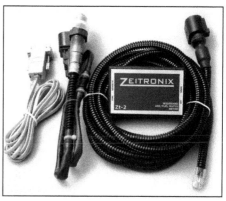

The Zeitronix wideband and its display are a great tuning tool. One simple display (they offer a couple of different types) can show boost, EGT, air/fuel ratio, and other parameters. It helps that Zeitronix has been a longtime supporter of the Mitsubishi tuning community.

number. Excessive EGTs can result in melted plugs, pistons, and worse.

Finally, unless you're tuning with an ECU that has a MAP sensor, you will need a boost gauge to keep tabs on manifold pressure. A useful MAP sensor can be added to the stock ECU with DSMlink, and most aftermarket ECUs contain one. An integrated MAP sensor is the best way to monitor boost, but an analog gauge mounted in view of the drive or dyno operator will work just as well. Choose a good gauge, though, since knowing boost pressure is critical to getting the best performance from a turbocharged motor.

CYLINDER HEADS, CAMSHAFTS
AND VALVETRAIN

In the past few chapters we've pushed home the idea that in a turbo engine, most of the engine's power comes from the turbo. That's true, but a significant part of your performance and engine flexibility is going to come from a good-performing cylinder head and camshafts.

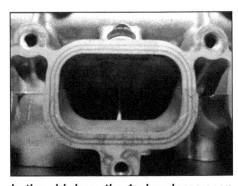

In the old days, the 1g head was seen as the ticket to big power because it has large, straight intake ports. It's still a good part for high-RPM-based power production, but not the best for midrange torque and a wide powerband.

A lot has been made of the differences in the DSM world between the 1g head and the 2g head. Sure the ports are different, but they're very similar otherwise, and there are many DSMs out there making big power with both the "small-port" 2g head as well as the "big-port" 1g head.

The design of the 2g intake port is actually better in most ways than the 1g port, especially for a street or road racing engine that has to pull from different RPM. For one thing,

the raised position of the 2g port gives the air a straighter shot at the back of the intake valve.

Air doesn't like to go around corners, so the straighter the port, the better the airflow, even when the cross-sectional area is smaller, as it is with the 2g port. Sure, the absolute flow numbers may be lower for the smaller port, but the velocity of the air is higher and will tend to help fill the combustion chamber at moderate engine speeds. In theory this will give the 2g head an advantage at anything

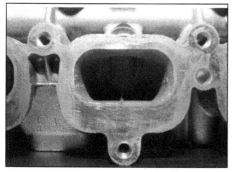

At first glance, the 2g DSM head would appear to be worse for power production than the 1g head. The ports are smaller, and angled up more. But appearances can be deceiving. This head can make as much power as a 1g head in all but the biggest-turbo applications, and on the street it's a star.

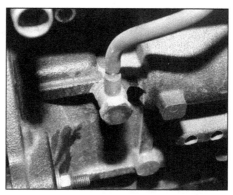

The Evo IX head has a separate oil feed for the MIVEC system from the original oil pressure switch location on the back of the block. To use a MIVEC head on an earlier block, the oil port has to be machined, and of course the IX head gasket, oil filter adapter, and ECU have to be used.

The MIVEC intake cam has a smaller drive end that works with the internally oiled MIVEC intake cam pulley. The other end drives a special CAS that tells the ECU what the current intake cam timing is, so that it can vary the duty cycle of the solenoid to achieve the targeted lobe center timing. Only MIVEC parts can be used with this cam (as well as the head).

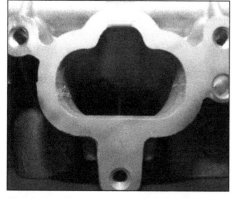

The later Evo ports are basically the same as the 2g DSM intake ports with an added bump for the injector spray to clear the port wall. They're raised the same amount and have the same area, but of course they don't have injector bosses in the port.

The oil feed passes through a solenoid on the head. At high RPM, the ECU activates the solenoid to pressurize a gallery, which in turn sends oil pressure to the front cam bearing and into the hub of the cam pulley. The pulley then rotates slightly to retard the intake cam's timing.

under full throttle, high-rpm running. In practice either of the 4G63t intake port shapes will work well for street and most race applications.

An interesting fact about 4G63t heads is that the final Evo VIII and IX engines have intake ports very similar to the DSM 2g ports. Mitsubishi knows how to build a port for maximum flexibility, and in this case the later 4G63t port is just that.

The USA-market Evo models have the ultimate 4G63t heads. Evo VIII heads have ports that are similar to the 2g DSM ports. The exhaust ports are also the same as DSM ports, but the manifolds position the reversed turbo in the wrong place for a DSM. The Evo VIII valvetrain can be used in a DSM, however.

The Evo IX added the most significant variant of 4G63t head. The head is not much of a redesign on the surface—the valves and valvesprings are the same as the Evo VIII head, and can still interchange with the DSM heads. The big change is in the intake cam and oil passages. Mitsubishi added variable valve timing to the 4G63t for the first time. While

MIVEC (Mitsubishi's name for VVT) was available in other cars, it finally came to the 4G63t for the last year of the engine's existence.

MIVEC works by having a cam drive pulley that can move to advance or retard the intake cam. This opens up the breathing at high RPM without compromising the engine's idle and low-speed performance. Later MIVEC systems, like the ones on the Evo X's 4B11 engine turn both cams to create different cam timing, unlike the single-cam system in the Evo IX.

Non-turbo 1g DOHC 4G63t heads are essentially the same as the turbo heads, although they often have different plumbing for water and oil than turbo heads, and smaller combustion chambers (43 cc vs. 47 cc). The stock 4G64 SOHC head is much worse than any DOHC head for performance. While it has the same exhaust bolt pattern and similar exhaust port size as the 4G63t, it's not the best head for performance use because of its poor airflow capabilities. In general, a 2.3 or 2.4-liter engine is going to want more port flow than a 2.0 liter engine, so think more about porting and bigger cams as part of your build. Any head will work, but more airflow will pay relatively bigger dividends on a larger engine.

Head Swapping

One of the neat things about the Mitsubishi Sirius family of engines is that every head can bolt onto every block. The head bolt pattern and bore spacing are the same. That means you could put a Evo IX MIVEC head onto a 1.6-liter 4G61 in a Mirage if you wanted to. Of course the accessories, manifolds, and head orientation will be all over the place, but the basic arrangement of bolts and bore centers is the same. Starion

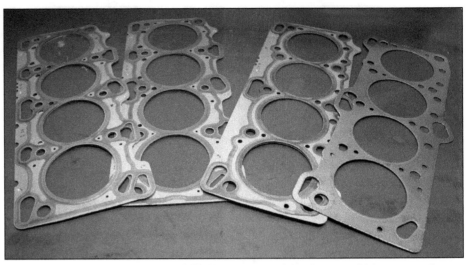

One look at this comparison of the various head gaskets and it's pretty clear that all of the 4G63t engines share the same head bolt pattern and most of the same oil and coolant passages. From left to right, the gaskets are Evo IX MLS, Evo VIII MLS, 2g MLS, and 1g composite.

drivers take advantage of this when building "wide block" 4G64 truck engines with 4G63t heads, and DSM builders have been taking advantage of it from the first time someone swapped a 6-bolt 1g head onto a 7-bolt 2g block, or the first time someone built a turbo 2.4-liter engine with a DOHC head.

The most common swaps are 1g head on a 2g block, and 1g or 2g head onto a 4G64 block. It's only when you get into the really high power range that you should consider swapping heads or manifolds around, and even then it's probably not worth the effort. The 1g head swap requires solving the CAS/firing order problem by using a 1997–up CAS or doing some serious welding on the end of the head to use the 1994–1996 CAS (see Chapter 6, for more on this) and using the 1g water plumbing (radiator hoses, thermostat housing, etc.).

It also requires some modification of your intercooler plumbing, and you have to use the 1g intake manifold modified to take the 2g MAP sensor, so it's not the magic bullet that some have made it out to be.

On the same topic, if you want to use a 7-bolt head on a 6-bolt engine, make sure to drill out the head for the 12-mm studs (versus the 7-bolt's 11-mm studs). Don't worry about the slightly larger holes going the other way (when putting a 6-bolt head on a 7-bolt engine). Dowel pins (not the bolts) hold the head in position and the head bolts are "dry" so it's not as bad as it sounds.

We talk more about 4G64 motors in Chapter 8, but any matching (left-versus right-mounted) 4G63t head will work for a 2.4-liter hybrid motor. You will have to use a different timing belt since the stock 4G63t belt will be too short, unless you happen to find one of the very rare DOHC versions of the 4G64. We'll cover all of this in more detail below.

Most people start with a SOHC 4G64 block from a Galant or Eclipse. The SOHC uses a completely different system for tensioning the cam belt, one that does not include a hydraulic tensioner or idler pulley. The timing covers are (of course) different, and some minor parts are different. To do a DOHC swap onto a 4D64 block, use

The DOHC Sirius engines have a hydraulic tensioner that isn't present on the SOHC engines, which use a different system for driving the cams. A general rule is: Regardless of the block, keep cam drive parts, oil pumps/front covers, and manifolds with the head you're using.

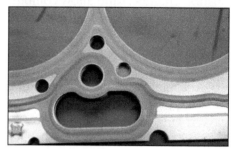

The Evo IX MIVEC head uses a different head gasket with restrictors in it to reduce oil flow to the head, since the MIVEC system provides a separate oil supply. Make sure to use the MIVEC head gasket if you're doing a MIVEC head swap, and don't use it on a "regular" Evo VIII head.

This shows the DSM head gasket laid over the Evo head gasket. It clearly shows the different oil drain holes between the early DSM-type heads and blocks and the later 7-bolt heads and blocks. You'll have to plug these holes if you use a DSM head on a late 4G64 block.

all of your existing 4G63t front parts with a longer timing belt. Also make sure to remove the tensioner spring anchor stud from the 2.4-liter block so that it clears the timing belt properly.

Late 7-bolt SOHC 4G64 blocks actually have one more hitch to using a DSM (6- or 7-bolt) head: The

blocks have an additional five oil drain holes around the outside edge that must be plugged. Evo heads bolt on without such a problem since the later heads have the same oil drain holes on both sides of the block.

Unfortunately Evo heads cannot be swapped onto 1g and 2g engines, because of the reversed ports and manifolds. The heads can be physically bolted on if you have a way to adjust valve timing and provide for the missing oil drain holes (easily accomplished by using an Evo or late 4G64 block), but all of your parts would end up on the wrong side of the engine, and the turbo would be somewhere under the dash. There are a few road race cars that have done this swap in order to get a front-mounted intake manifold, and it has uses in RWD swaps.

The Evo IX head can be swapped onto an Evo VIII or 4G64 block (or any other Sirius block) with some

machine work. Additional wires have to be added for the MIVEC actuator and secondary CAS if you're swapping the resulting hybrid into a car other than an Evo IX.

For all motors other than the MIVEC Evo IX, make sure you use a head gasket that matches the block (not the head) you plan to use, since there are some oil and coolant passage differences between some of the 2.0 and 2.4-liter blocks. In addition, the stock-type head gaskets don't have fire rings large enough in diameter for 2.4-liter engines. The fire ring cannot hang over the edge of the bore.

Head Gaskets

Unfortunately, blown head gaskets are a frequent occurrence with heavily boosted 4G63t engines. Head gaskets usually fail because of a mechanical problem or poor installation—they rarely fail "just because."

The most common source of head gasket failure is the same as the most common source of piston and bearing failure, namely detonation or pinging. The extreme forces of detonation pound the edge of the fire ring (the steel ring around each combustion chamber) and allow combustion gasses to get underneath and erode the ring.

Overheating can also cause a blown head gasket as the head expands and contracts more than usual and "scrubs" the surface of the gasket. Once this happens, the same process of erosion and damage occurs. Finally, simple combustion pressure from a very high-output engine running very high boost levels (between 25 and 30 psi, depending on the turbo), can cause the head to lift slightly from the block surface and allow combustion gasses to get under the head gasket and into the cooling system, blowing out all of your coolant and causing the engine to overheat.

Sometimes a head gasket will blow only partially, blowing out at high RPM and full boost, while running normally and showing no signs of gasket leakage at all other times. This kind of problem is hard to diagnose at first, but it rarely stays this way. Within a few miles or a few runs, the gasket usually fails completely and the symptoms are the same as any other failed head gasket.

The easiest way to test for a blown head gasket is to look for signs of either oil or combustion gasses in the coolant, since these are the most common leaks. Look into your coolant reservoir for signs of milky deposits caused by oil. Smell the coolant reservoir with the engine idling; if it smells like exhaust, the head gasket is leaking from the combustion chamber. Failing this, check for coolant in your oil in the form of

After teardown, the ideal thing to do is have the head and block meticulously refinished to a very smooth, flat finish. Of course this is not always possible, since it requires removing and disassembling the short-block to achieve perfection.

the same white gunk on the dipstick. As a confirmation, use a hydrocarbon-detecting probe in the water neck or a combustion gas tester kit on the coolant.

If you've blown a head gasket, try to figure out why it happened before fixing it. Did the engine overheat? Does it need more radiator? A bigger cooling fan? Were the head bolts fresh and properly torqued? Make sure that you know or at least have a good idea why the gasket died so that you can avoid the problem in the future.

With a stock-finish block and a freshly machined head, your choices for a new head gasket are somewhat limited, since some of the strongest gaskets don't like to seal against a less-than-perfect surface.

Measure the head for warpage. If the head is warped at all, make sure your machinist checks for warping or misalignment in the cam bearings. You want them to maintain a perfectly flat relationship to each other, which cannot be achieved if the head is warped more than a few thousandths and then milled flat.

Gasket Types

The original Mitsubishi gasket in your engine is either a graphite composite gasket (commonly used on early DSM engines) or an MLS (multi-layer steel) head gasket. The composite gaskets have a steel fire ring around the combustion chamber and a thick, soft surface that seals well to irregular machining and old head gasket marks. The Mitsubishi service parts replacement gasket is the best-known and best-quality version of this gasket, available from your nearest Mitsubishi dealer or speed shop.

MLS gaskets are a very good choice for a high-powered street engine because of this toughness, but

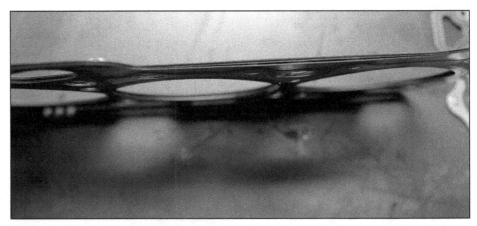

MLS gaskets are much tougher because they are made from a stack of thin steel shims riveted together at the corners. They can withstand more detonation and cylinder pressure before failing.

Mitsubishi supplies an MLS head gasket under part number MD349347 for the DSM engines. It's a good, inexpensive choice for a high-performance street rebuild, as shown by many 400+ hp street cars still running the stock head gasket and bolts.

A new version of the Cometic MLS gasket uses a fire ring separate from the rest of the gasket. This is a good solution for chronic head gasket problems if you don't want to cut for an O-ring.

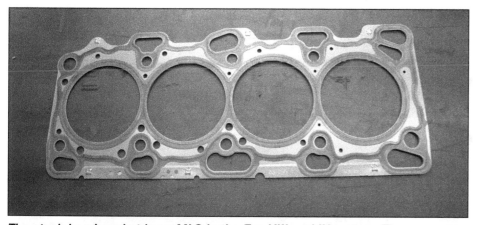

The stock head gasket is an MLS in the Evo VIII and IX motors. These are very good gaskets. Stick with stock unless you want to experiment and you're prepared to pull the head if the gasket doesn't seal right.

The fire ring part of the multi-piece Cometic gasket has to be replaced with each head removal, but the rest of the gasket can be reused on the same head and block.

they are not as forgiving as old-fashioned graphite composite gaskets. They require a good surface preparation of both the block and head. Both mating surfaces must be very smooth. Always use one of the more forgiving head gasket materials, such as the stock Mitsubishi composite gasket, if you cannot resurface both the head and the block.

Cometic and Fel-Pro both make MLS gaskets, but the Cometic gasket in particular, requires a very fine surface finish, much finer than the Mitsubishi or Fel-Pro gaskets need, to seal well. The Cometic gasket is reusable on the same block and head, as long as it is not damaged.

HKS, Cosworth, Toda, and several other companies also produce very good quality MLS gaskets that can be reused if treated properly. Surface preparation is key to getting any gasket to seal well, and for MLS gaskets it goes double. Sometimes a quick squirt of copper sealant on both sides of the gasket will aid in sealing, but this should not be necessary if the surfaces are properly prepared.

Copper head gaskets are cut from a single thin sheet of copper, which makes them very strong and available in different thickness. This allows the engine builder to set up the exact combustion chamber volume and squish clearance that is desired. Copper head gaskets are very strong and can be reused as long as you don't resurface the head or block. However, they are difficult to seal perfectly. Copper gaskets normally should be used with sealant such as Hylomar or copper gasket sealant to eliminate oil and water leaks and sometimes they need to be anmaled (heated and cooled slowly).

They usually also require that the block or head be prepared for metal

O-rings are a great idea on paper but expensive and difficult to make work in practice. Some people have had success running into the 10-second range with OEM head gaskets and copper O-rings, but they are in the minority. One reason to think twice about O-rings is that if a head or block is cut for one, it's impossible to get a normal gasket to seal in that motor.

O-rings around each combustion chamber. A shallow, thin groove is cut all the way around the head, the block, or both. A stainless steel or copper wire ring is then carefully laid into this groove with the ends butted together. Talk to your machinist about O-ringing if you want to try it. He or she's probably tried several different ways and may have a favorite.

Head Bolts and Studs

As we hinted above, improper installation is one of the biggest causes of head gasket failure. The factory has a required pattern for torquing down the head bolts, and this should be adhered to. If you use the stock hardware, stick to the stock torque specifications, and re-torque the head after a few hours of running if you use a composite gasket. 6-bolt engines have strong, 12-mm studs that can be reused indefinitely, while those in 7-bolt engines are 11 mm and "torque to yield," which means they stretch when installed and should not be reused. Many people

When you get your block back from the machine shop, but before you install the head studs and gasket, make sure you have both of the head dowel pins. They're very important to prevent the head from moving around and damaging the head gasket.

do reuse them in moderate performance builds, but it's a false economy since the failure of your head studs means replacing the head gasket.

As you get above the 350-hp (big 16G) range of airflow and cylinder pressure, you'll want to start thinking about replacing the stock head bolts with something stronger. ARP is the de facto standard in aftermarket head studs. Although there are other

brands available, none has achieved the market share of ARP, and none are as easy to find. They're popular for a reason; they are reasonably priced and very good quality. A set of "standard" ARP bolts is more or less standard practice for most 4G63t rebuilders. They offer slightly more available clamping load than the stock bolts but they also do a better job of spreading clamping forces around the stud because the nuts and washers are larger diameter than the stockers.

The ultimate in head sealing is the ARP L19-series studs. They shouldn't be tightened any more than the standard A1 studs, but they are more resistant to stretching and losing their clamping force with very high cylinder pressures. For really high-powered "dyno queens," these are the best choice, but most DSM and Evo builders won't need the extra insurance that they provide.

Note that if you want to be able to remove the head in the car, you will have to trim the front motor mount bracket on a 1g DSM engine to use studs, but otherwise they are

If you use studs, torque them to 85–90 ft-lb and no further. Studs can stretch if they are over tightened, and this destroys their clamping ability. Torque studs in stages, making sure that the final jump is larger than the others to overcome static friction in the threads. A good torque sequence would be 10–30–55–90 ft-lb, for example.

Always use A/RP anti-seize lubricant on the top threads, but leave the block threads perfectly clean and dry, and the studs screwed in only finger tight. If you don't use anti-seize, your torque numbers will be inconsistent and possibly meaningless.

Dollar for dollar, not much will beat a professional port and polish job. You can make some improvements yourself, but don't expect too much from a home porting job on a modern engine like the 4G63t.

direct replacements for the factory bolts. Re-torque your head bolts after a few hours (or a few thousand miles) of running. Don't loosen them, just put a torque wrench on them and make sure the torque has not changed from when you assembled the motor. After two or three uses, even ARP studs will stretch beyond reuse. Check your studs for stretching if you pull the motor apart, and throw them away if you find any that have loosened from stretch.

Port Flow

The stock 4G63t ports are very good, but there is still room for improvement. Better flowing intake ports will mean you can make the same power with less boost. On the exhaust side, more flow will help spin up your turbo faster and lower the boost threshold, but not by much. That's because the turbo is the biggest restriction to exhaust flow, not the valves and ports.

Porting is a messy, dirty job, and to do it right you will need to spend some money on the right equipment. The shape of each port is actually

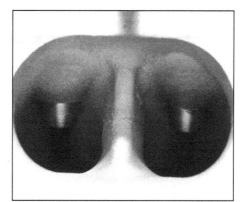

Simple things you can do to help airflow include "knife-edging" the divider in the intake and exhaust ports. The stock divider is wider than necessary, and not very aerodynamic.

more important for power production than the size of the port. It's also more important than the surface finish. A perfectly polished port has minimal advantage over a rough port, even on the exhaust side. On the intake side it can even be detrimental to flow.

You can do minor porting (with noticeable gains) on your own head, but go slowly and don't expect too much without a flow bench to try dozens of different designs. The Mitsubishi ports are very good—not

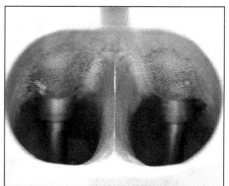

The ported head has a thinned and sharpened divider between the valves, but notice that the size and shape of the port has not changed. Even this professionally ported Evo head has stock-sized exhaust ports, since the stock ports have plenty of flow.

many heads flow as well as a stock 4G63t. It's easy to ruin this good flow, for example, by hogging your 2g intake ports out to 1g size.

Limit your modifications to lightly cleaning up the port entrance to match the manifold. Don't go too wild; just clean the port up within the gasket outline. Don't feel you have to go all the way out to the gasket outline, just eliminate any step between the intake manifold and

On the valve end of the port, just blend the valve seat into the wall of the port. Again take as little material off as possible, just enough to smooth out the step you can see here between the seat and port.

head. Make sure that the head ports are larger than the manifold ports, and make sure that on the exhaust side the step goes in the opposite direction, if there is one. Don't change the shape of the turn into the port, and don't remove too much material from the guide boss, since both areas are important for flow. If you want to really make power, pay someone with a flow bench to port your head. Successful head porters spend many hours on each head in order to develop the best shapes for power. It takes dedication, and a lot of ruined heads, to get improvements in airflow and velocity from heads like these.

One of the most important revolutions in cylinder porting in the last 10 years is the explosion of CNC porting. The idea is a simple one. Instead of spending hours hand porting intake and exhaust ports to match a not-very-precise template, porters simply program a CNC machining center to duplicate any port shape you want. The devil is in the details, of course, and the port design and programming are critical.

Since port shapes are well developed, at least in theory, each head will flow as well as the master head. There are a couple of companies

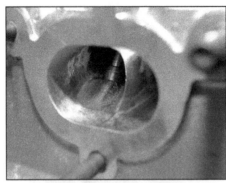

A CNC ported cylinder head has many advantages over a hand-ported head. Each port is exactly the same, so per-cylinder airflow and power variation is reduced to basically nil.

offering CNC-ported heads for the 4G63t (Cosworth is the leader in the market), and they can be a very good, but expensive, choice for getting more airflow.

Combustion Chamber

All 4G63t heads have a very similar combustion chamber shape, from the first 1g DSMs all the way to the Evo VIII. The Evo IX had a longer-reach plug and a slightly different combustion chamber shape designed to get the spark deeper into the head and provide more detonation resistance.

Professional porters always balance the combustion chambers to each other to equalize compression ratio between the cylinders. The nominal 4G63t combustion chamber volume is 47 cc. When you build your motor, don't rely on published compression ratios; measure the volume of each combustion chamber and piston dome yourself and calculate the ratio from the actual bore and stroke you are using.

If you're doing your own porting, just clean up the combustion chamber, removing bumps and rough surfaces without changing the shape. Carefully mark the head gasket fire rings on the head and do not go outside this line for any reason to avoid affecting head gasket sealing. Polish the chamber as much as possible to

The 4G63t has a nice combustion chamber by modern standards—compact with little surface area and a dished piston. The four valves mean that valve shrouding (when the head and cylinder get too close to the edge of the valve and reduce flow) is not really a problem.

Each engine porter will have his or her own ideas about combustion chamber shape. Most cut a little around the valves and remove casting marks, but rarely do porters modify the 4G63t combustion chamber. This one's just been opened a little to accommodate larger valves.

Aftermarket guides are not a necessity for most head rebuilds if the stock iron guides (below) are in good shape. Replacement guides, on the other hand, are ready to go out of the box with the proper stem-to-guide clearance, and they will last for a long time.

keep carbon from building up and encouraging detonation.

Make sure to do any porting before having any machine work done. A slip of the die grinder is much easier to repair during the machining of the deck surface and valve seats rather than after.

Valvetrain

All 4G63t engines have the same valve size and stem diameter (6.6 mm), as well as the same valve length (109.5 mm intake and 109.7 mm exhaust). There are significant differences in the retainers and springs between the Evo and DSM engines, but they are interchangeable as a package (retainer, springs, locks). All DSM and USA-market Evo engines have the same 34-mm-diameter intake valves and 30.5-mm-diameter exhaust valves.

Evo exhaust valves are sodium-filled for durability, though. The sodium transfers heat from the exposed head of the valve to the tip and thus to the head and oil, reducing valve head temperatures.

For most street applications, the stock valves are fine. They're a good

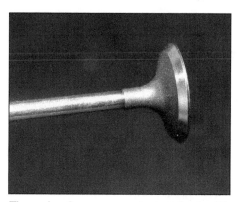

The valve faces are cut at a single 45-degree angle, standard practice these days. Most stock valves can be reused with a quick regrind; they are made of very good materials.

size for street-turbo power levels, and going to larger valves can be expensive. Larger, aftermarket valves require machine work to the valve seats and combustion chamber to use properly, especially on the exhaust side.

On the other hand, aftermarket valves are usually stainless steel (sometimes with chrome-plated stems) and can last longer than the stockers, as well as have head shapes that are more conducive to good airflow. Notice the gradual "tulip" shape of a good aftermarket valve and compare it to the hard corners of

the stock parts. Aftermarket valves usually also have undercut stems, which means that the stem takes up less room behind the valve and allows more room for air to flow into the combustion chamber.

Aftermarket valves are often lighter, which helps because, in the valvetrain (as with the bottom end of the engine) light weight is vitally important. The lighter the valve parts, the less likely they are to float, or stay off the seat at high revs due to inertia that overcomes the valvesprings. This goes for the springs, retainers and locks, too.

Guides and Seats

Mitsubishi stock valve guides are cast iron, which is strong, conducts heat moderately well, and is cheap. Cast iron is not as good as other materials at transferring heat to the head and oil, and it must be run with large stem-to-guide clearance. This encourages the valve to wander around the perimeter of the seat and not seal as well, and it also allows a little oil to get past the guide.

Aftermarket guides are generally bronze alloy, which has better self-lubricating properties than cast iron,

Notice that there is more room for larger valves on the exhaust side of the combustion chamber, so if money is tight consider building a head with stock intake valves and larger exhaust valves.

DSM springs (on the left) are decent; they're good to less than 11 mm of lift at the valve, which covers many aftermarket cams. The Evo springs are good to similar lift levels, but they are a more modern design that's good to a higher RPM at about the same lift.

and can therefore be run with tighter guide-to-stem clearance. This helps valve seal and reduces oil getting into the combustion chamber.

4G63t valve seats are made from very hard steel pressed into the aluminum head and generally don't require much service. Usually, a used head won't even need to have the seats cut to be serviceable. A good machine shop can open them up for significantly larger valves, but not all shops will be willing to cut them so large without replacement.

On the subject of the actual valve seat, seat thickness is critical for airflow—the thinner the seat the better the airflow, although seat and valve life decreases the narrower the seats get. The stock 4G63t has a 3-angle valve job to help flow at low valve lifts. The valve seals against a standard 45-degree angle, but the other angles serve to smooth the airflow into and out of the valve seat. Make sure that your machinist also cuts the valve seats with at least three angles. Three-or-more-angle valve jobs are commonly performed as part of a performance head rebuild, but verify that yours will be cut this way.

Springs, Retainers & Locks

The valvesprings in the DSM motors are interchangeable, and not particularly special. They're strong enough to prevent valve float below 8,000 rpm or so, although they will tend to float at a lower RPM as they age. The retainers are plenty strong enough, but heavy, which encourages valve float at high RPM.

Evo VIII and IX springs are much improved. They are "beehive" shaped, that is, narrower at the top. So-called beehive springs are better because they can give high-RPM performance similar to a dual or stiffer stock spring, while reducing spring pressure on the cam. In addition, the narrower top of the spring makes for a smaller, and thus lighter, retainer. They are anodized aluminum and as light as almost anything from the aftermarket.

You won't gain much by spinning a street motor to 9,000 rpm, even if you can shift at that speed. This is even less of an issue with stroker motors. Most 2.4-liter engines, even with all forged internals, won't be happy spinning at high RPM for other reasons.

Don't forget to install these little shims under the valvesprings. Without them the springs will dig into the aluminum head. Stock and aftermarket springs require spring seats like these, but check that they fit your springs if you plan to use them.

If you're building a DSM, try to get your hands on some Evo exhaust valves, retainers, and valvesprings. The retainers and springs will allow a little more revs and cam lift, while the exhaust valves are sodium-filled for reliability. The intake valves are the same.

Always use your spring manufacturer's recommended retainers and locks. Not all locks are machined at the same angle and mismatched locks and retainers is a recipe for a dropped valve. The same goes for retainers—use new retainers with new valvesprings, and keep them matched to the right keepers.

It's fine to reuse springs and retainers in a stock rebuild, but check that the springs have enough seat pressure to reuse, and they are not collapsed. The factory shop manual doesn't give pressure specifications but it does say DSM springs should be discarded if they are under 47.3 mm tall. Evo springs should be taller than 49.4 mm. Make sure retainers match the springs, and check used ones carefully for damage or wear.

Don't change the valvesprings to aftermarket unless you want to run a big camshaft or increase the rev limit. Stronger valvesprings with lighter retainers are very good to help prevent valve float at high RPM, but aren't necessary on most 4G63t engines because the engine and turbo run out of breath at an RPM lower than valve float.

Hydraulic lash adjusters can be noisy if they are not properly bled during cam installation, and the precise passage inside the HLA can become plugged with oil sludge or bits of grit. The 3g HLAs have a larger hole in the tip and are less prone to plugging.

Hydraulic Lash Adjusters & Followers

The 4G63t, like most modern engines, has low-friction roller cam followers and hydraulic lash adjusters (HLAs) that take up clearance between the follower and cam while allowing the valves to close

Evo and DSM followers are manufactured differently; DSM followers are forged and Evo followers are stamped, but they are interchangeable. The Evo followers are a little lighter, but not enough to make them necessary.

completely. The 4G63t's stock roller followers have a ratio of 1.7:1, so for each millimeter of cam lift, the valve raises 1.7 mm of its seat.

The followers rarely wear out, and since they run on hardened rollers, they don't have to be matched to a particular cam. Inspect them carefully for grooving or roughness, but they can usually be reused for many thousands of miles.

HLAs do a good job of taking up wear in the various parts involved in opening the valve, from the cam to followers and the valve, which reduces noise. In general, that's the main job of an HLA; to make the valvetrain quiet for many years, and eliminate the maintenance of valve adjustments. Some people call them "lifters," after cam followers in an OHV pushrod engine, but that's not an accurate description.

If you're completely rebuilding a head, replace all of the HLAs. Use the latest 3g adjusters. They're so-called because they were originally installed in the USA in the 3g Eclipse non-

turbo 4g64 engine, but they are also the same as those found in the Evo VIII and IX—MD377054 is the latest part number. High-quality aftermarket lifters are also acceptable. Make sure that your valvetrain geometry is set up correctly, and clean and bleed the HLAs properly during installation following the procedure in the shop manual.

HLAs have some compromises for performance use. Since they are designed to operate in a narrow range of valvetrain geometries, rebuilt heads often have trouble with noisy valves or valves that hang open. In addition, at sustained high engine speeds, HLAs can "bleed down" or collapse from the constant battering and high engine oil pressure. This will reduce lift at the valve and cause the car to run badly. Next time you hear a 4G63t on a dyno, listen for valve clatter after each pull. This is a sign of the HLAs deflating.

For most engine builds, this isn't a problem, particularly on the street. High-RPM race engines are the only ones that will be able to benefit from aftermarket solid adjusters. Of course, it takes a lot more time to set them up, since each valve must be adjusted separately by removing the cam and adjusting the solid adjusters, but it won't have to be performed very often.

Valvetrain Geometry

Usually your machine shop will set up the valves and springs for you, but checking it yourself is cheap insurance. It helps to use a machinist that's familiar with the 4G63t so that they won't make any basic mistakes on your head, such as cutting the valves too deep, thinning the seats too much, or messing up the installed heights of the valves and springs.

The first and most important parameter you have to watch out for is valve installed height. Check the height of each valve above the head before you disassemble it, and ask your machinist to match the installed height to this number. That may mean removing a little bit of material from each valve tip if you reuse stock valves, or cutting aftermarket valves slightly to increase installed height.

The second important factor in valvetrain geometry is spring installed height. Measure the height from the spring seat (don't forget to replace the metal seat when you reassemble your head) with a caliper, and adjust if necessary with shims or by machining the spring seat. If you're using aftermarket springs, make sure that you get proper installed height specifications from the manufacturer. Stock 4G63t valvesprings have an installed height of 40 mm (around 1.575 inches).

When you put the head together, run the cams through a few rotations with the head on the bench (spaced up so that the valves can open) to

The relationship of the valve tip, cam follower, HLA, and camshaft is important to build a quiet, reliable head that's happy spinning at 8,000 rpm. Any deviation from the right relationship could result in valves that hang open, among other troubles.

check clearances. Make sure that the springs are not binding. The coils of each spring should not touch causing the spring to jam up.

Finally, make sure that there is sufficient clearance between the valvespring retainer and the top of the stem seal with your chosen cams. You should have at least .030 inch of clearance. Check using a solid lifter (see the pictures for an example of a solid lifter), and a soft "checker" spring. If the clearance is close, check all of the valves, and have your machinist adjust any that are out of spec by cutting the valve deeper (not recommended) or shortening the valveguide. Don't forget to check valve-to-piston clearance on any unproven build (you should have at least .045 inch).

Camshafts

If there was no such thing as inertia or momentum, the intake valves would open just as the piston reaches top dead center (TDC) after the

exhaust stroke, and close right when the piston reaches bottom dead center (BDC) on the compression stroke. In the real world, it takes time, and degrees of crankshaft rotation, to get as much air as possible into and out of the cylinder. In fact, the intake cam opens the valves before the piston starts going down the cylinder to allow time for all the air and fuel to get into the cylinder.

At the same time, the valves stay open past the bottom of the intake stroke going into the compression stroke. Conversely, the exhaust valves open before the bottom of the power stroke, and close after the piston reaches the top of the exhaust stroke. This means there is some overlap in valve opening at top dead center between the exhaust and intake strokes.

The number of crankshaft degrees on either side of the intake stroke and exhaust stroke are the critical period for cam timing and thus engine breathing and power band. The longer the intake valve stays open, the more time the intake charge will have to rush into the cylinder at high engine speeds, while the longer the exhaust stays open, the more time the exhaust gasses will have to get out of the cylinder.

The problem with this is that at low speeds, the more duration you have in crank degrees, the more overlap you have, and the more the idle quality and low-speed performance will suffer. There are other effects of shorter and longer open durations, but the shifting of the power band is the most important. Stock 4G63t cams have between 244 and 260 degrees of intake timing, and between 244 and 256 degrees of exhaust timing depending on the year and model.

The other parameter that cams control is the lift of the valve from its seat. Lift is important for good breathing, but it is limited by geometry. With short cam timing, you can't get the valve open fast enough without damaging the valvetrain, and the pistons and other valves might be in the way at the ends of the valve open period. The longer the duration, the more time is available to lift the valve fully.

If you are going to compare two cams, be very careful with the manufacturers' published numbers. Duration is either theoretical (so-called advertised duration) or the duration at some small amount of lift, such as .050 inch or 1 mm. Advertised duration is the amount of crank degrees that the cam lobe takes up, assuming no lash in the valvetrain. Of course in the first part of that period the valve won't be open enough to even begin flowing air, which is why the second duration is the more realistic number. It's difficult to compare two cams very closely on paper, which is unfortunate. At the very least, make sure you're comparing them both at advertised or measured (.050 inch or 1 mm) lift.

In a way, cams are like turbos—they are a compromise in airflow and power potential between low-speed torque (like a small, fast-spooling turbo) and high-rpm power (like a massive drag turbo). Like everything else in the stock 4G63t, the cams are oriented towards good idle, low emissions, and reasonable power. The DSM engine's high-RPM airflow capacity and horsepower production is really choked by the stock cams. The Evo VIII and IX cams are much better, but there is still plenty of room for improvement in high-RPM breathing with some compromise in every other area.

Race motors can use much more aggressive cams, but the stock Mitsubishi cams are pretty good for street use with the stock turbo. It's not until you upgrade the turbo and do all the usual bolt-ons that you should consider swapping cams, but they become more important when you start to change the engine configuration drastically.

For example, strokers like bigger (longer duration) cams. The bigger the engine the bigger the cam you can run without compromise. You should run a bigger cam than stock on a 2.3 or 2.4-liter engine to get more air into your larger cylinders at high speed. The stock cams will quickly run out of breath above 5,000 rpm on a 2.3-liter engine. A ported head can make use of bigger cams, too, since high-rpm flow won't be limited by the head.

Incidentally, the problem with the 4G63t isn't really one of lift; it's one of duration. Lift isn't as important with 4-valve heads like the 4G63t, since most flow occurs at fairly low lift, and there just isn't much room to increase lift without having valve-to-piston contact. The most radical 4G63t cams don't go over 12 mm (.480 inch) of lift, while stock Evo cams are 10 mm.

There are a lot of nuances to picking cams that really won't fit in this book, but there are some generalities. In theory, you can have any cam lobe profile ground onto any cam you want, but in practice very few people will go to that trouble. Most of us will be choosing among the available aftermarket cams, most of which range from 262 to 280 degrees total duration. If you have an Evo IX, you really have one choice in intake cams—the Cosworth cams.

For a street car, you want a cam with fairly conservative timing and lift under about 10.5 mm so you can use stock-type valvesprings. If your car will be expected to pass a tailpipe emissions test, keep your cam timing

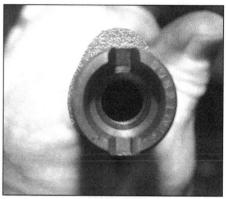

The Evo (bottom) has a CAS on the exhaust cam instead of intake like a DSM (top), and the slot is in line with the lobe instead of 90 degrees to it. To use an Evo cam in a DSM you can machine a CAS slot in the intake cam. This is one cheap way to get some upgraded cams into your DSM motor.

conservative. Less than 210–215 degrees of effective open time is the magic number, more than that and the overlap period between intake and exhaust starts to get too long for clean idle emissions.

Turbo engines don't need big cams anyway. Huge cams don't work well with high boost, as they usually result in an explosive power band and high EGTs. Keep overlap to a minimum. You want as little overlap as possible because lots of overlap allows pressurized exhaust gasses to reverse into the combustion chamber. This heats up the charge and reduces efficiency at high boost, right where you want efficiency to be the highest.

Intake duration should be more conservative than exhaust duration on a turbo engine. Many people combine "race" exhaust cams like the HKS 272s with "street" intake cams like HKS 264s to get the best combination of good idle, flexibility, and high-RPM breathing. Evo IX builders often keep the stock intake cam and run an HKS 272 on the exhaust, which is an excellent combination.

Use 1992–1994 1g manual transmission cams for the best timing and lift of any DSM cams (the ones you want are marked "D" on the intake and "C" on the exhaust). Evo cams are 260 intake/256 degrees exhaust duration, and are great cams for the 4G63t. They can be phisically installed in a DSM head although the CAS runs off the exhaust cam instead of the intake, and is phased 90 degrees differently than the DSM cams.

Cam Drive

All 4G63t engines use the same timing belt: an 18-mm-wide round-toothed belt with 168 teeth, part number MD326059. DOHC 4g64 engines need a longer timing belt to make up for the difference in deck height (part number MD182292), so make sure you order the correct belt for an engine build. The DOHC–SOHC swap makes the timing marks on the cam pulleys have a different relationship to the crankshaft timing mark if you don't compensate for it. The DOHC 4g64 cam pulleys have the correct timing marks to work with the 4G63t crank pulley, but you don't need to use them.

If you use 4G63t pulleys, make new timing marks one-half tooth to the right (advanced) of the existing marks. Remember which marks to use when you assemble the timing belt.

With adjustable cam gears on the 4G63t's cams, you have control over lobe separation or the degrees of crank rotation between the center of intake and exhaust lobe. Play with lobe centers on the dyno after building the engine to achieve the best power curve. This can pay off in a wider power band and smoother idle than just installing your cams "straight up."

Don't be tempted to use an aftermarket tensioner or idler pulleys; most of them are not up to the quality level of the stock parts. Always replace the balance shaft belt when you do a timing belt, too—the results of a failed belt are much greater than the minimal cost of buying a new belt and installation.

Cam Timing Sequence

For all discussions of piston and valve movement, the "0" point from which all valve opening and closing events are measured is when the piston is at the top of the cylinder, referred to as top dead center or TDC. TDC is easiest to find with a piston stop like the one Robert Garcia's holding.

Find TDC by setting the piston stop. Allow the piston to come nearly to the top of the cylinder but not all the way, and then bolt on a degree wheel and turn both directions to find where the crank stops. Adjust the degree wheel so that the stops are both at the same point before TDC (BTDC) and after TDC (ATDC).

Now that the degree wheel is centered around TDC, you can remove the piston stop and begin the process of finding the lobe centers of the intake and exhaust cams.

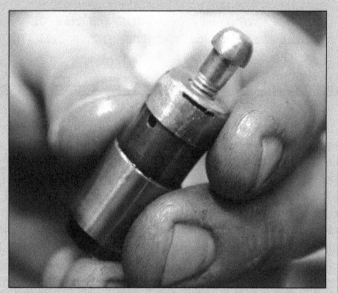

To find the lobe centers, you need a way of measuring lift at the valve, and you need a way of preventing the HLAs from collapsing. It's easy enough to build a simple mechanical HLA like this one from an old HLA; a scrap bolt and a nut.

Cam Timing Sequence *CONTINUED*

Use a dial indicator on the valve to find the point at which the valve opens 1.0 mm (.040 inch). Then find the point at which the valve closes to .040 open. Add the two together, add 180 degrees, and divide by 2 to get your cam centerline. If your intake valve opens .040 inch at 10-degrees before TDC, and closes 20-degrees after BDC, your centerline is 105 degrees (10 plus 20 plus 180 divided by 2).

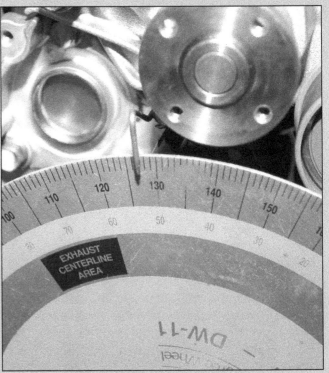

Compare the degrees shown on the degree wheel to your cam card and adjust your cam gears to change the centerline. Spin the engine through again and check once more. Repeat the process for the exhaust cam.

Whatever cams and pulleys you use, think about running adjustable cam pulleys to get the best balance of power and drivability. Increasing lobe separation can help make big cams more streetable, although it will take away some of their breathing ability on top. Tightening up the lobe centers increases overlap and can make idle worse, but possibly help with high-RPM breathing. The bigger (more duration) cams get, the more critical lobe centers become, and the more power can be found in playing with cam timing.

On the other hand, if you don't plan to spend time adjusting cam timing on the dyno or track, and you haven't milled the block or head, then adjustable cam gears are not necessary. If you've milled the head, cam timing will change because the tensioner pulley for the cam belt is on the right side of the cam belt. Both cams will be retarded slightly from their intended crank timing. This will actually help high-RPM breathing somewhat as the exhaust valves close later than intended, but there will be a matching decrease in low-RPM performance.

As a time-saving note, when you change the timing belt or cams, just use cable ties to lock the belt to the cam pulleys, which will eliminate losing timing while tension is taken off the belt.

It's always a good idea to properly dial in any aftermarket cam that you install, because tolerances could add up to a cam that doesn't open and close when you think it does. If you plan to install adjustable cam gears, this goes double. Sure you could install the gears straight up and play with the timing until the engine runs the way you want it to, but knowing where your lobe centers are and where you want them to be is very helpful when setting up the engine on a dyno.

THE BOTTOM END

The 4G63t's tough bottom end has features that have become increasingly rare in these days of carefully optimized production engines, increasing emissions regulations, and drastic weight-saving programs.

Of course, like any motor, the 4G63t has its weaknesses. Some motors have thrust bearing problems with no clear solution, and the balance shafts are not optimal for performance use. In addition, the engine's smallish 2-liter displacement is an obstacle to really big streetable power. The stock pistons and rods are strong, but they are only production parts.

These weak spots can be overcome with the use of careful assembly and the right combinations of stock and aftermarket parts. Good hardware, pistons, and quality machine work will give you an engine that can run with some of the best production engines ever unleashed on the aftermarket. Even better, the 4G63t's displacement can be easily stretched to 2.3 or 2.4 liters with related engine parts for better off-boost performance.

The 4G63t's block is one of the engine's great strengths. It is rather low-tech, has lots of material in all the right places, and is nearly unbreakable absent abuse. Subaru, Honda, and Nissan tuners can only dream of the power levels achievable with a mildly built 4G63t short block and the right supporting parts.

This is a 1G DSM block with the oil pan removed. Notice the sturdy main bearing girdles connecting the front and rear pairs of main bearing caps. These "suitcase handles" add significant strength to the turbo engine's block.

Stock 4G63t Bottom Ends

The 6-Bolt Engine: Early 1g DSM and Galant VR-4

The earliest 4G63t engine founded the reputation of Mitsubishi turbo engines as powerful and bulletproof. It had massive, thick connecting rods; wide, heavy rod and main bearings; a (ridiculously low) 7.8:1 compression ratio; huge intake ports and manifold; and a big turbo. The crankshaft was forged steel, nitrided for surface strength.

The first and second main bearing caps were paired, as were the fourth and fifth mains. The pistons were cast, but thick, and had low, wide rings for strength. Oil squirters sprayed cooling oil at the bottoms of the piston domes for reliability. Two balancer shafts combated annoying harmonic vibrations.

This was an engine designed for high power output with lots of boost. Its off-boost performance was lethargic, but when the boost and revs came on, watch out! It's a tough engine but suffers from lots of rotating weight. Each following version of the 4G63t was lighter, revved more freely, and was designed with a close eye on economy and off-boost response.

7-Bolt Engine: Late 1g, All 2g DSM and Non-USA Evo I–IIIs

All 2g DSMs and later 1g DSMs came with a variant of the 7-bolt engine. It is a good strong engine with some notable improvements over the 6-bolt engine. First, the 2g version has an almost 1.0 point higher compression ratio (8.5:1), for better off-boost performance.

The piston rings, while the same width, were redesigned and seal better than the earlier rings. The rod and main bearings were narrowed to decrease frictional drag in the engine, and the lighter rods with smaller bolts help the engine rev more freely, though they are a little weaker. Larger wrist pins help to better distribute loads between the piston and rod. Later engines received an improved center thrust bearing design that is self-aligning. The oil squirters were changed slightly, but remained.

This engine can be less forgiving of tuning mistakes than the earlier engine but is capable of making just as much power in most combinations, often with less boost than a 6-bolt engine. It has gotten a poor reputation because of thrust bearing problems (see "crank walk," detailed later) and weaker pistons and rod bearings.

The 6-bolt and 7-bolt 4G63t engines take their names for the number of bolts holding the flywheel to the end of the crankshaft. There are hundreds of detail changes between the two engines; about all they share is bore and stroke, a head bolt pattern, and bellhousing bolt pattern.

With the introduction of the 7-bolt 4G63, the main bearings went from having separate paired caps to being one large casting, which makes the bottom end more dimensionally solid and stronger. This design was carried over with very few changes to the final Evo VIII and IX.

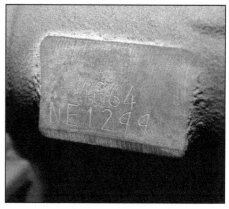

The stock 4G64 engine doesn't have much power potential, so it's not a common swap. The block, however, is a good starting point for a torquey 2.4-liter street motor.

"Reversed" 7-Bolt Engines: Evo VIII and IX

After the last DSM in 1999, the 4G63t took a hiatus from the USA market. It didn't become petrified in other markets, however. Mitsubishi continued to tweak the basic 7-bolt formula to make the engine stronger, more powerful, and lighter. The pistons got narrower rings, and redesigned rods. Extra oil drains were added to the block. The crankshaft gained nitride hardening again, and the oil squirters reverted to a type similar to the old 6-bolt engine's squirters. Overall, though, the bottom end changed very little from the debut of the 7-bolt in 1994 until the final Evo IX in 2006. The Evo IX got some detail improvements, but the engine carried over mostly unchanged except for the addition of external oil lines for the MIVEC head.

The 4G64/G4CS

The 2.4-liter 4G64 motor is basically a bored (to 86.5 mm) and stroked (to 100 mm) version of the 6-bolt 4G63t. It has the same big, heavy rods, and a forged nitrided crankshaft, though it has weaker cast pistons. The G4CS is its Hyundai counterpart.

The 4G64 is also available as a 7-bolt motor from the mid 1990s on. Late 7-bolt 4G64 motors have more drain holes on the block surface than the 6-bolt and 7-bolt DSM engines, but the same number as the Evo 7-bolt 4G63t. Late 7-bolt 4G64

The 4G64 engine block casting is essentially the same as the 4G63t casting but 6-mm taller to accommodate the longer stroke, and without the oil gallery machining needed to run oil squirters. Early 4G64 engines also have separate main caps. This makes the main bearings less stable than those in a turbo engine, but they are still very strong.

More Displacement

Displacement is the area of the bore times the stroke. All else being equal, the bigger the displacement the more power your engine will make. It's the old "no replacement for displacement" saying made modern. Sure, a turbo tends to equalize smaller engines with larger ones, but add a turbo to the larger engine and watch the fun really start.

Starting with the stock 4G63t at 1,997 cc (85-mm bore and 88-mm stroke), and the G4cs/4G64 at 2,353 cc (86.5-mm bore and 100-mm stroke), adding in the widely available oversize pistons from 85 to 87.5 mm and crankshafts at 88, 100, and 102 mm, you end up with many possible engine displacements.

There are two limiting factors for displacement. The first is dimensional. There is only so far you can stroke and bore a particular engine within the dimensions of the block. The second is mechanical. The longer an engine's stroke, the lower its rev limit must be. That's because each piston has to accelerate from a dead stop at the bottom of the bore, to full speed halfway up the bore, and then decelerate to a complete stop again at the top.

The longer the stroke, the greater the forces on the piston. To simplify the calculations, average piston speed is generally used to determine the forces on the piston. Piston speed is generally given in feet per minute (FPM), which is calculated as 2 x RPM x stroke in feet.

The maximum piston speed for a stock cast piston and connecting rod is somewhere in the 4,000–4,500 fpm range. Much above this requires very expensive parts to achieve reliably, but modern 600-cc motorcycle engines have no problem hitting 5,500 fpm regularly. The 88-mm stroke of a stock 4G63t gives you 4,000 fpm at around 7,000 rpm and 4,500 fpm at 7,700 rpm, which isn't very high. With light-weight rods and really nice forged pistons, 4,500 fpm isn't a problem and 5,000 fpm is a more realistic ceiling on piston speed, but these are all very general rules. With a stock crankshaft, 5,000 fpm comes at 8,600 rpm, which is plenty high enough for all but the very largest turbos.

What happens when you stroke the engine to 100 mm? The results are surprising; you get 5,000 fpm at just above 7,500 rpm. Don't even think of using cast pistons with a high-strung 4G64 crank, because 4,000 fpm arrives at a very low 6,200 rpm!

But don't get hung up on high engine speeds. If you can swing the cost, a stroker motor or full 2.4-liter motor makes an almost perfect street powerplant. The low-RPM boost you'll get from a large-displacement engine, and the resulting quicker boost onset, will be more useful on a daily basis than the high-rpm rev potential you might get from a short-stroke 2.0-liter engine.

As for the dimensional limits, in the case of the 4G63t, they're very well known. A little under 2.5 liters (2,456 cc, to be exact) is the maximum that a stock 4G63t block can be taken. This assumes the maximum available bore (87.5 mm) and the maximum available stroke (102 mm).

But before you start cutting, consider that an increase of stroke will net you more displacement than an over-bore. For example, going from an 85 mm bore stock-stroke 4G63t to the largest 87.5 mm overbore nets you only 2,116 cc, or about 6-percent more displacement. On the other hand, switching crankshafts to the largest aftermarket 102 mm crank with the stock pistons gives 2,270 cc, or nearly 14 percent more displacement! Both displacement increases require new pistons, so clearly the longer stroke is a more cost-effective way to gain displacement. The larger bore will have other effects on engine breathing at high RPM, but read the following block section about larger bores and engine strength before committing to a huge overbore.

---

engines have joined main caps just like their 4G63t counterparts.

With the exception of the 1994 Galant GS, all USA-market 4G64 and G4CS motors were shipped with a SOHC, 2-valve per cylinder head. Almost every change listed above for the 4G63t also applies to the 4G64/G4CS—the 7-bolt engines have smaller rods and lighter pistons, and the crankshafts are not nitrided. There is a "reversed" 4G64 that is installed in 2000–2005 Galant and Eclipse models, which can be used to build a 2.4-liter Evo motor.

Machining a Block

All of the 4G blocks—the 6-bolt and 7-bolt 4G63t, and 4G64/G4cs—are slightly different. The 6-bolt and 7-bolt blocks require different crank, rods, bearings, oil pans, and front covers. The same goes for the 2.0-liter and 2.4-liter blocks. The 6-mm deck height means that the proper internal parts have to be used. The non-turbo blocks are identical to the turbo blocks with the exception of the oil squirters.

For a high-performance rebuild, start with a rebore too. You can run a stock bore if it's within spec, but many people have the block bored to start with virgin metal for best ring seal and a slight displacement increase. The best machinists use a torque plate to simulate the loads and distortion put into the engine block by a torqued-down cylinder head. A torque plate should always be used with the head gasket and head bolts or studs you plan to run in the finished motor.

The maximum possible 4G64 bore is 87.5mm using aftermarket pistons. This isn't a particularly reliable combination though; 85–86 mm bore engines will be much stronger over the long haul because the wall thickness at 87.5 mm is so thin that it impacts ring seal. In extreme cases, engines this big can break out the side of the cylinder bore under high boost. On the street it's not likely, but poor ring seal and head gasket can be.

For the street, keep the bore under 86 mm or keep boost down with larger bores. If you're going to use a larger bore than 86 mm, use a 4G64 block—not because it has thicker walls (it doesn't) but because it will allow you to use longer rods with "stroker" pistons for a more favorable rod ratio. If you do go the 4G64 block route, keep boost and cylinder pressure reasonable.

Before the block is bored, have the deck surface trued up slightly. Don't take off too much—you don't want to impact the compression ratio or decrease your piston-head clearance. Remove just enough to pro-

Before you go out to the largest size, remember that the 4G63t has only a little room to increase the bore. Some 4G64 blocks have slightly thicker sleeves, but in general they are no better than 4G63t blocks. In addition, the bores in a 4G64-based engine are very close together—too close for optimal head gasket sealing.

Have the wall thickness checked with a sonic measuring tool if you want to know exactly how much room you have to increase the bore, but generally 87 mm is fine with either block (as shown here). Notice how close the pistons are in this shot compared to the 85.5-mm bore shown above.

duce a polished surface that will bond well with the head gasket.

Have your machinist check the alignment of the main bearing bores—any misalignment or shift should be eliminated by align boring the block. The only way to ensure that the main bearing bores will be properly sized and aligned is by align-boring them.

To use the 4G64 crank in a 6-bolt 4G63t block, the suitcase handles have to be clearanced for the longer stroke of the 4G64's crankshaft. Use an angle or bench grinder to remove the material. It's not particularly critical, just make sure you remove enough to give you at least 1/8-inch clearance.

Even if the main bearing bores are in perfect alignment, they should be align bored if you plan to use aftermarket main studs. These require greater torque than the stock studs, and this will distort the block and caps slightly.

Notice how much clearance there is between the 2-liter connecting rod cap and the suitcase handles. The stock engine has about 1/4 inch or more.

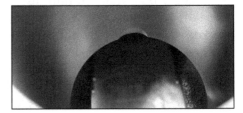

You may need to clearance the sides of the bore with some combinations of stroke and connecting rod brands. Wait until you have your connecting rods before attempting this, because clearance can vary depending on the bolts and rods used.

Seven-bolt engine main caps have room for a stroker crank, but require some clearancing. Just be careful not to break through the casting. Do the clearancing before sending the block out for machining if possible. The chips and dirt that the grinding process produces will be cleaned out when the machinist hot tanks the block.

Balance Shaft Removal

All 4-cylinder engines lack inherent balance in one plane. They vibrate at particular RPMs, depending on stroke, piston, rod weight, and other factors. Since the 1970s at least, Mitsubishi has installed balancer shafts to combat these harmonic vibrations.

A balancer shaft is basically a steel shaft that rotates at faster-than-crank speed. It has a carefully designed imbalance that counteracts that in the engine it is designed for. Normally two are installed, to cancel the rocking motion produced by a 4-cylinder engine.

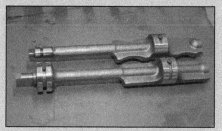

Because of the way 4-cylinders are imbalanced, there is no way to cancel all of the vibrations produced by the engine, except just those that occur in a particular direction and at a particular range of engine speeds. Also, balancer shafts are carefully designed to combat vibration occurring in one particular engine—if you change pistons or rods to a different part, or stroke the engine, or bore it larger, the balancer shafts become less effective.

Balancer shafts have a few detractions in a high-performance 4G63t. First, they increase the effective rotating weight of the bottom end. They are run at double the crank speed, which means they have a large impact on acceleration. Second, the shafts have plain insert bearings like a camshaft, and one is driven by the oil pump. If a bearing fails, it has the potential to damage the pump and cause catastrophic engine failure.

Also, the other balance shaft's drive belt is very close to the main cam timing belt. If a problem occurs with the balance shaft system, the belt can break. A broken balance shaft belt will very likely foul the cam belt and cause it to skip teeth and go out of time. The result is bent valves and an expensive teardown.

Most 4G63t engine builders, including Robert Garcia, remove the balance shafts whenever they build an engine. It's cheap, and the benefits of eliminating them remove any chance of their failure. It's not hard if the engine is out of the car. It can be done with the engine in place but is not recommended because there isn't as much room to work. To remove them properly, at least the front cover and oil pump must be taken off for access.

There are a couple of ways of disabling the system. The easiest (and least elegant) is to remove the independent shaft completely, and cut off the business end of the pump-driven shaft. If you're attempting this in the car despite the warning above, just leave the independent shaft in place after removing the pump-driven shaft and pulleys; there may not be enough room to get the old bearings out and rotate them.

No matter what method is used to eliminate the balance shafts, their oil supply must also be blocked off to preserve oil pressure. The pump-driven balance shaft is oiled by radial holes

The best way to disable the balance shaft system involves replacing the pump-driven shaft with a Mitsubishi-made stub shaft produced for the balance-shaft-less 1.6-liter 4G motor in the Mirage turbo. The stub shaft is shown on the right. On the left is the standard oil pump drive gear.

Some have eliminated the pump-driven shaft's balancing properties by turning it down on a lathe. The assumption is leaving the rest of the shaft in place will help stabilize the driven oil pump gear. Real-world testing has not shown this method to offer any benefit over the standard way of eliminating the balance shafts, so it is probably not worth the effort to duplicate.

under the oil pump gear, and a hole drilled the length of the shaft. If you simply cut off the oil pump stub end of the shaft, you will have to tap the center of the shaft for a longer retaining bolt that will seal off the feed holes in the shaft. The short eliminator shaft is

Balance Shaft Removal *CONTINUED*

not drilled for oil supply so it does not require any preparation beyond ordinary deburring and cleaning.

There are a few more issues that need to be dealt with. The first is the hole that the balancer shaft belt tensioner was threaded into; it is an open hole into the front engine cover, so make sure to plug it with a short 8-mm bolt and a bit of Loctite. The second is the crankshaft pulley for the balancer

belt. You can just leave it in place, or, if you want to save every possible ounce from the crankshaft, you can turn down the old pulley to create a spacer. There is the 40 mm hole in the front cover where the independent shaft seal was. Plug it with a 40 mm freeze plug.

Balance Shaft Elimination Parts:
- Front balance shaft front bearing, Mitsubishi PN MD040597

- Front balance shaft rear bearing, Mitsubishi PN MD103722
- Balance shaft plug, Mitsubishi PN MD092785
- Oil pump drive shaft, Mitsubishi PN MD098626
  (You might be able to reuse the old bearings, but new ones will fit tighter and are cheap insurance if you have trouble getting the old ones out without damage.)

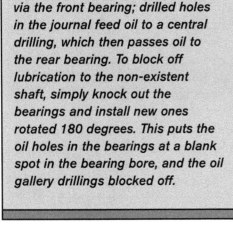

The belt-driven shaft is lubricated via the front bearing; drilled holes in the journal feed oil to a central drilling, which then passes oil to the rear bearing. To block off lubrication to the non-existent shaft, simply knock out the bearings and install new ones rotated 180 degrees. This puts the oil holes in the bearings at a blank spot in the bearing bore, and the oil gallery drillings blocked off.

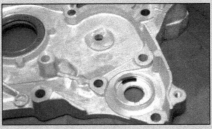

After removing the independent shaft, plug the large hole in the cover that formerly contained the seal with a 40-mm freeze plug. Use the Mitsubishi plug for the best fit—the metal around the seal is very thin and can break if you use too much force or a poorly fitting plug.

Seal the plug with a coating of Loctite to prevent it from coming loose, and use JB Weld or similar metal repair epoxy on the inside to make it permanent—a leak here would dump out a lot of oil very quickly.

The Sirius engines have a front-mounted oil pump driven off of the cam belt. The oil pump should be replaced any time you rebuild the engine. Even when it's within specifications, a used pump should be tossed because it's not an expensive part and it is very critical to the engine's health.

Oiling System

All of the stock pumps should be able to maintain 15 psi at hot idle, and will top out at somewhere above 75 psi at high RPM. Don't depend on the factory gauge, it's neither very accurate nor very quick to respond. Low oil pressure isn't usually caused

Early 6-bolt engines have straight oil pump gears, but later 7-bolt engines have teeth that are cut at an angle. The straight-cut oil pumps whine a little more than the angled-cut teeth, but they are functionally the same. Both work very well.

by pump wear, instead it's the result of wear in the crank bearings that allows oil to escape more quickly than the pump can shove it in. If you have a low oil pressure situation, pull the pan and check the bearings before you consider pulling the oil pump.

Clean your oil galleries very thoroughly while you have the motor apart, including the crankshaft oil passages. Oiling passages start at the main bearing journals and are drilled

Always use OEM Mitsubishi pumps, and always replace the front cover at the same time, since it forms the other half of the oil pump's wear surface. A proper rebuild includes oil pump gears, housing, and engine front cover.

through the crank counterweights to each rod journal. Oil from the main bearings gets to the rod journals through intersecting drilled passages, the ends of which are plugged with a soft metal slug.

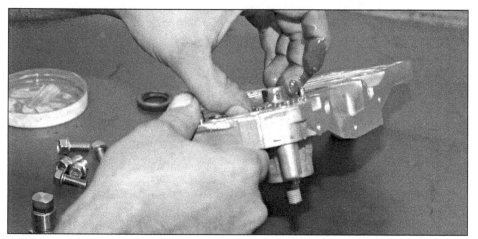

Always pack the oil pump with oil or assembly lube before installation. This will help the engine build oil pressure as soon as possible after the build. Assembly lube has the advantage that it won't drain back down while the motor's sitting there ready to go in the car.

A good trick to make sure the oil pump drive gear doesn't come loose is to put a spot of weld on the nut. Sure, it makes it impossible to remove later, but the cost of a new drive gear and nut is a small price to pay for the peace of mind you'll have knowing it can never come loose. Don't forget the SOHC engines (4G64 7-bolt) use a different driven pulley and gear than the DOHC engines.

For the ultimate in cleanliness, have your machinist remove the soft metal plugs from the oil passages and tap them for threaded plugs. Before installing the threaded plugs, your machinist will clean the oil passages until they are spotless and tighten the threaded plugs (with Loctite and peened threads to keep them from ever coming loose).

Oil squirters are a good idea for an endurance engine, but a lot of street and drag racing motors have been built without them. You should be at least running forged pistons to run without squirters, since stock pistons don't have the crown strength to handle high temperatures.

Over thousands of miles of use, dirt and foreign particles in the oil get flung out into the crank's oil passages and build up. Handling the crankshaft out of the engine and any machine work that is done to it can all result in chips, abrasive, or dirt getting into the oiling passages. At a minimum, the oil passages in any used crankshaft should be thoroughly cleaned with a narrow bottle brush and lots of solvent.

The 2.4-liter blocks (and the non-turbo 4G63 blocks from 1g NTs) do not have piston oil squirters in the main oil gallery. The cast bosses are there, however, and can be drilled and tapped into the main oil gallery if you want to use squirters with your non-turbo block. A good machinist can thread the bosses, but many people believe that they are not necessary. Drag racing and street motors, in particular, can probably survive without them since runs are short, and piston crowns shouldn't be subjected to constant high temperatures. If you're building a road race or rally

Install the squirters early in your build after cleaning out the block. Make sure you mark them (and all critical engine fasteners) in case you put the engine down for a day or two before finishing it.

racing motor, try to incorporate the oil squirters if you can. They improve reliability in any engine.

The stock DSM oil cooler is a water-cooled matrix mounted on the oil filter housing. They're a good part, but can leak after many years from corrosion between the coolant passages and oil. In addition, they

add thermal load to the stock radiator, which in a DSM is known for its marginal size.

A high-performance engine used in competition should be converted to use an external air-oil cooler. The Evo VIII and IX housing can't be used to provide oil ports because it's mounted differently, but the Evo III one can for 7-bolt engines. 6-bolt blocks should use the 1990 DSM filter adapter with external oil ports.

Pistons, Pins and Rings

Six-bolt engines have a 21-mm wrist pin and 7-bolt engines have a 22-mm pin but, otherwise, 4G63t pistons are similar with the exception of valve cutouts. All of them have a deck height (from the center of the wrist pin to the top of the piston) of 35 mm.

Compression Ratio

Generally 4G63t compression is determined by the piston dome,

A turbo engine should really have a large air-to-oil cooler. Even Mitsubishi understood this, that's why the supplied all Evos after the Evo III with a large, front-mounted cooler. The stock Evo cooler is very good; consider it for your DSM if you are pushing the car hard.

The massive aftermarket unit under the bumper of this Evo shows careful attention to detail in the mounting and plumbing. Use the right filter housing and plumbing to make sure the install doesn't leak.

All stock 4G63t pistons are dished, with different dishes to get the different compression ratios. Lancer Evo IV to IX pistons have reversed valve reliefs; the intake and exhaust reliefs are cut opposite those on the non-rotated engines. From left to right these are 1g 6-bolt, 2g 7-bolt, and Evo VIII pistons.

because the various combustion chambers are all around the same size. The domes are flatter and less dished in the high-compression 2g and Evo engines, for example. Choose your compression ratio like you do anything else—based on the expected use and lifespan of your engine. Remember that lower compression ratios (down to 8.0:1 or so) are easier to tune, since you can run more boost and timing without worrying so much about knock and EGT.

Higher compression ratios (CRs) with a little less boost and a little less timing can make more power, at the expense of tuning precision. A low-compression engine will be more forgiving of a tune that's not spot-on.

If you live in California or another 91-octane state, keep your compression ratio low—8.5:1 is the max for 91 octane with a turbo. If you can get 93 octane with regularity, you can run an additional .5 or so of CR. At the same time, if you are running a turbo that likes really high pressure ratios for maximum efficiency (like a 54-trim T3), keep your CR low to use lots of boost; if your turbo likes moderate boost (like a 16G), you can run a higher CR.

The stock pistons have a thick top surface, which makes them able to withstand lots of boost. Stock pistons are cast, which is a good method for making pistons because cast pistons can be run with very tight bore clearances, for one thing. For another, though, cast pistons have a tendency to break because they are not as strong, nor as ductile (able to bend) as forged pistons. That means that it doesn't take much detonation to cause them to crack.

The weakest point of any piston is the top ring land. The thin ring of piston material is fragile because it's not supported, and detonation can cause it to crack and break off. This will result in poor ring seal, oil burning, and eventually engine damage. Other failures include cracked pin bosses, the result of over-revving, but they are much less common on a street motor that rarely sees the high side of 7,000 rpm.

Forged pistons are stronger and more ductile than cast pistons, making them the natural choice for a high-performance build. The more powerful the motor, the more important forged pistons become. High cylinder pressures caused by lots of boost and lots of revs is hard on pistons, so big horsepower engines need all the help they can get.

The best reason to upgrade your pistons on a mild motor is to get a larger tuning window. A combination of high EGT and detonation will destroy pistons quickly, so a little extra durability is always a good idea. This Evo VIII piston suffers from the classic signs of a pinched top ring land—once blow-by gets past the top ring it acts like a blow-torch, eating away at the side of the piston.

Forged pistons are much tougher than cast pistons in material, as well as design. The distance from the top ring to the top of the piston is critical for durability in a turbo engine, and many aftermarket pistons have rings that are lowered compared to stock.

Forged pistons do have some downsides. Because of the way they are made and the alloy they are forged from, some forged pistons need more piston-to-cylinder wall clearance than cast pistons. This can result in noise, especially when the engine is cold and the forged pistons rock in the bore. This is generally called piston slap; it's not the end of the world, but it can be annoying to live with.

The additional 12-mm stroke of a 4G64 crankshaft in a 4G63t block requires a change in either block height or rod length to prevent the pistons from popping out of the block. Higher-pin stroker pistons are designed to use a 100-mm crank with stock 4G63t block and connecting rods.

When you pick an aftermarket piston for your build, pay attention to the weight of the pistons as well as their strength and dimensions. Piston weight is nearly as important as piston ultimate strength and ductility. As we discussed in the section on stroke, heavier pistons are subjected to higher stresses from over-revving. The lighter the piston the less force acts on it, and the higher the rev limit that can be safely used. A lighter piston of the same strength is a much longer-lasting piston than a heavy one.

Rings

Early 4G63t engines have a 1.2-mm top ring, a 1.5-mm second ring, and a 3.0-mm oil ring, with several different designs depending on the year of the engine. The Evo engines got a 1.2-mm second ring and a 2.0-mm oil ring for less ring drag and better sealing.

The stock Evo rings are very good. The top ring is made from steel, which makes it strong and flexible. The second ring is ductile cast iron, which is another premium ring material. The rest of the stock 4G63t rings are also made from ductile iron, and they are probably superior to the most common aftermarket rings. Many street engines have run well over 200,000 miles without needing new rings.

Some engine builders swear by Total Seal brand gapless rings, which, in theory, should provide nearly perfect bore sealing. In practice, the ring gap isn't the biggest contributor to blow-by, and the difference might not be as much as you expect. However, they are very good quality rings, and there is some evidence to suggest that gapless rings can give a slight power boost, especially when new.

Always hone the block to match the type of ring being used. Check with your ring manufacturer if you're

These cylinder walls have been honed on a power hone to the perfect crosshatch finish. Talk with your machinist about the pistons and rings you'll be using in your motor because different pistons have different honing requirements and some manufacturers recommend a coarser cut than others.

Wash the bore with dish soap and water to get rid of any abrasive grit from the boring and honing operations. Use a soft brush and really scrub the bores down thoroughly. Blow the whole thing off with air of you have it, or towel it off with blue rags if you don't.

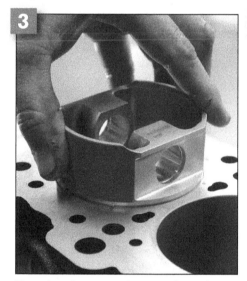

To properly gap a ring, push each one down the bore with a piston to make sure that it is square. This can be tricky with oil ring support rails if you're using a piston with a 3-piece oil ring, but the accuracy of your gapping job depends on it.

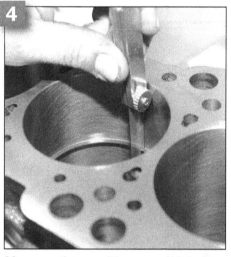

Measure the resulting gap. If it is less than .016–.018 inch (for a top ring), grind the end of the ring with a small file until the gap increases. The second ring should be gapped a little wider—.018 to .024 is a good number to shoot for.

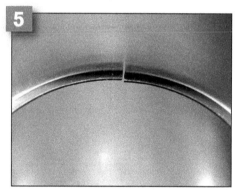

Always deburr each ring as you finish it, using a small fine oilstone. Keep track of which rings go in which cylinder so that they will be fitted to the cylinder they will be used in.

using aftermarket rings. For Mitsubishi chrome rings, a very fine surface finish is required. A poor bore finish is the most common source of ring sealing problems.

Most performance aftermarket rings are "gap to fit," meaning they are manufactured slightly oversized and must be cut down to fit the cylinder. Ring gapping is a slow, laborious process, but the results are worth it. Rings that are gapped too small can butt together and break, while those that fit sloppily can cause excessive blow-by and poor break-in. The best ring gap tool is a dedicated electric or hand ring-gap grinder, but a small file works well instead.

Connecting Rods

All stock 4G63/4G64 rods are the same 150 mm in length. The small and big ends are different depending on the year of your motor, and there are significant differences in the design of the beam, and thus the rod's strength. There are differences in fasteners as well. The 6-bolt rods

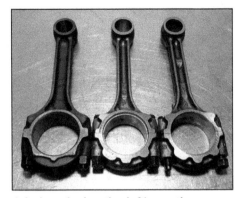

6-bolt rods (on the left) are the biggest of the 4G63t rods. The 7-bolt (center) are strong enough for most uses, though they have an undeserved reputation for being weak. The Evo VIII and IX connecting rods (right) are the strongest 7-bolt. The bolts are the weak point of all 7-bolt rods, however.

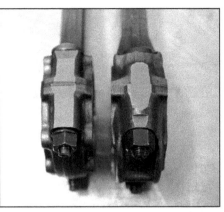

The 7-bolt big-end bearing (on the left) has a nominal width of 26 mm, while a 6-bolt (right) is 28-mm wide—a difference of 2 mm, or .080 inch. You can use a 6-bolt rod in a 7-bolt engine with some machine work, but it probably isn't worth it.

rather than in the rod and eliminates the need to capture the rod in the piston. Most aftermarket rods, on the other hand, are designed for a floating rod that is captured by circlips in the piston.

Reconditioned Stock Rods

For years DSM engine builders have used forged-steel 6-bolt rods because they're stronger and larger. These rods are definitely the toughest of the stock DSM rods, but they're also the heaviest, which is a significant tradeoff. The 6-bolt rods have to be precisely narrowed by 1 mm on each side to run them on a 7-bolt crank.

The small-end should be re-sized and machined too. Stock 6-bolt rods will need to be bored out for 7-bolt wrist pins, which are 1 mm larger. If you're planning on using them with stock 2g 7-bolt pistons, make sure that the small end is narrowed as well, since it is wider on the 6-bolt rods. This isn't an issue with aftermarket pistons.

Some people convert the stock press-fit rods to a floating rod, but this requires complicated and expensive machine work. The cheap way is to simply hone the small end out until there is enough clearance for the pin to float. This makes the wrist pin bearing steel on steel, which is not good for extended high-RPM use. The proper way involves boring the small end of the rod oversize and pressing in a bronze bushing with the proper inner diameter (ID) for the wrist pin. Oil holes then have to be added to each rod to ensure that the pins don't seize on the new bearing.

As already mentioned, critical engine fasteners—rod bolts, main bolts, and head bolts—shouldn't be reused in a performance build, expecially torque-to-yeild fasteners. Bolts they should be fine in theory, but bolts that have been torqued to

High power levels require lots of boost and lots of revs, and that's hard on connecting rods. Aftermarket rods like these will give you hundreds of RPM of headroom. Notice that they're bushed for floating pins, too.

have gigantic 9-mm rod bolts that are heavy and stronger than necessary for the stock engine (and most modified engines). The 7-bolt engines, including the Evo VIII and IX, have 8 mm torque-to-yeild bolts that are both lighter and much weaker than the 6-bolt rod bolts.

The small and big ends of the 6-bolt and 7-bolt rods are different. Big end bearing width differs, although the bearings are the same diameter outside (48 mm) and inside (45 mm). The small end of the 6-bolt engines is 21 mm, while the 7-bolt motors got stronger 22-mm pins.

All Mitsubishi rods are designed to have the connecting rods pressed into them. This is simply so that all of the wear occurs at the piston

The 4G63t rods are some of the toughest production rods you'll find, and damage like this broken rod occurs only when there's an over-rev or oiling failure. Usually the bolts are the first parts to fail, so they should be replaced first.

within 80 percent of their ultimate tensile strength for the last 100,000 miles or 10 years are not to be trusted when thousands of dollars in parts and tuning are on the line. Replace the bolts as a matter of course, with ARP or other quality fasteners, if you can afford it.

With the new bolts installed and properly torqued, have the rod big ends re-sized. As rods are used, they tend to stretch a tiny bit, which reduces bearing crush in the big end. This will ensure that the bearings have enough crush or squish, because that is all that keeps them from working loose and spinning on the crank. Aftermarket bolts torqued to a higher specification cause another problem—this can pinch the rod bearings and cause failures from localized overheating.

Your machine shop will prevent both problems by milling down the flats of the rod cap very slightly and then honing the bore back to size. This honing process is why you always want to keep each rod with its mating cap.

Two other common operations performed on stock connecting rods are shot peening, the process of blasting steel shot at the surface of the rod to make it stronger and less likely to fail from repetitive stresses, and polishing. Stock rods are shot peened from the factory so it's not necessary to shot peen a set of stock rods. If you choose to polish them (which removes stress risers and makes them tougher yet) the polished surface should be shot peened to make it even stronger.

Aftermarket Rods

Aftermarket rods are commonly used in high-performance rebuilds, but they aren't really needed until you're trying to squeeze 400 or more horsepower out of your engine. A set of reconditioned stock 6-bolt or Lancer Evo rods with aftermarket bolts will be sufficient for most engine builds under 400 hp.

Most rod failures occur because of over-revving, so your RPM limit is really the determining factor in deciding if you need aftermarket rods. Of course rods can fail (by bending) from pinging or very high cylinder pressure, but this is much less likely than failure from over-revving.

All that said, there are two good reasons for switching to aftermarket rods on even a mild rebuild. First, a set of quality forged aftermarket rods (like the inexpensive imported Eagle rods) is only a little more expensive than buying new hardware and machining a good used set of stock rods. Instead of hand working and balancing a set of rods, buying new expensive hardware and shipping them all over the place for reconditioning, you simply open a box and bolt them in.

Second, aftermarket rods and pistons give you the option of changing the rod length (and therefore rod ratio), of your bottom end. Rod length can cause some difference in engine power output at moderate engine speeds, and has some impact on reliability.

Rod Ratio

An engine with a rod that's the same length center-to-center as the stroke is said to have a rod ratio of 1:1. There are factors that have a much larger influence on both midrange power and reliability, but rod ratio is one tool available to engine builders.

In general, the longer the rod, the lower the peak piston speed at any RPM. Of course the average piston speed will always stay the same since the stroke is the same, but the acceleration and deceleration at either end of the stroke will be less, so the forces on the piston are less than on a short rod. In addition, the longer rod reduces side forces on the cylinder wall, which helps reduce wear on the rings, piston, and bore. Assuming the same stroke and block height, a longer rod also requires a shorter, lighter piston, which helps the engine rev more freely.

Longer rods can also help engine breathing. Compared to a shorter rod, the piston in a longer rod engine spends more time on either side of top and bottom dead center, which gives more time for cylinder filling. A shorter rod, on the other hand, accelerates the piston away from TDC faster and can cause intake pressure to fall more quickly. This can help a low-flowing head at low and mid RPM, but since there's no free lunch, the accelerating rod also tends to give less time for cylinder filling at high RPM.

In general, long-rod engines are better for turbo engine use, but there are many counter-examples. With the stock 4G63t's 88-mm stroke and

Unfortunately there isn't enough room on the piston between the oil ring and pin to raise the pin more than 6 mm. This means the 162-mm rods cannot be used with a 4G63t block and 100-mm crankshaft, but they can be used with a 4G64 block.

150-mm rod, it has a decent, performance-oriented 1.7:1 rod ratio. In the 4G64, or a 4G63t with a 100-mm crank, the rod ratio is only 1.5:1.

Aftermarket 156- and 162-mm rods are available, allowing you to build an 88-mm stroke 4g63 with a rod ratio of 1.77:1, or even 1.84:1. The same rods with a 100-mm stroke crankshaft result in rod ratios of 1.56:1 or 1.62:1. So you can see that the longer stroke crankshaft does have some downsides in terms of rod ratio.

Only stock-length rods can be used with a 100-mm crankshaft in a 4G63t block because of the piston pin location. If you want to try longer rods, you will have to use a 4G64 block with a 100-mm crankshaft, or stick with the stock stroke in a 4G63t block.

Crankshafts

Stock 4G63t crankshafts have an 88-mm stroke, 57-mm diameter (nominal) main journals, and 45-mm rod journals. The width of the bearings varies between 6-bolt and 7-bolt engines; 7-bolt journals are narrower on both rod and main bearings. All stock Mitsubishi crankshafts are forged from alloy steel, and are shot peened before machining the bearing journals.

The 7-bolt cranks weigh less than the 6-bolt cranks, with smaller counterweights and lighter (narrower) rod bearings. The 6-bolt crankshafts have different main bearings and rear seal than 7-bolt cranks. Needless to say, the two types of crankshafts are not interchangeable.

Stroker Crankshafts

It just so happens that the 4G64/G4CS crankshaft uses the same main and rod bearing diameters and widths as the same-generation 4G63t. It is also forged steel like the 88-mm crankshaft. It also uses the same rods and piston pin height as the 4G63t, but the block is 6-mm taller to compensate for the 12-mm longer stroke.

An engine based on the 4g64 will require different valve timing gears and belt. In addition, the bigger bore on the 4G64 means it has weaker cylinder walls. The solution is to run the 100-mm crankshaft in the 2.0-liter block—a fairly easy swap.

If you plan to run a 4G64 crank in a 4G63t block, you will need to use "stroker" pistons to compensate. These have a pin that has been moved up 6 mm to compensate for the additional 12 mm of stroke. This gives them a 29-mm pin height versus the 35-mm pin height of a stock 4G63t engine.

Stock Crankshafts

If you are re-using the stock crankshaft, have the journals polished (which doesn't resize them but rather cleans up the finish to make the crankshaft kinder to bearings), and you should be good for as much power as the rest of your parts can handle.

If you do any machining on the crankshaft throws and cheeks, have the machined areas shot peened to increase fatigue strength and replace the shot peened surface you removed by grinding. Don't bother with lightening or grinding on the crank

A set of inexpensive aftermarket rods, like these Eagle rods, can be cheaper than having your stock rods resized and cleaned up. Add the fact that they're already bushed for floating pins, balanced, and available in different lengths and they make a convincing case for even a street rebuild.

Stock 100-mm stroke crankshafts are fine for a street motor, but a race motor should use an aftermarket 100-mm (or 102-mm) crankshaft like this one. The throws don't overlap as much as the 88-mm crankshaft's, so 4G64 cranks are prone to breakage in high-horsepower engines.

The stock 88-mm crank is heavier than necessary. If you want to squeeze the maximum power and acceleration out of your engine, have the stock crankshaft lightened. Limit such machining to an 88-mm stroke crankshaft, since 100-mm cranks are not as strong to begin with.

A good machinist will take significant weight off of the counterweights and the sides of the crank throws to reduce rotating weight, tapering the material removal like shown to reduce drag and oil churning in the crankcase.

throws for a street engine, you probably want maximum reliability rather than ultimate rev-ability.

Have the crank checked for straightness and straightened in a press if there is any bend noted, and then have it properly balanced. Most factory cranks and all aftermarket cranks will be balanced very precisely out of the box, but it never hurts to make sure you haven't been given a bad one. Your machinist will be able to check balance, but make sure to include the pulley you plan to use as well as your flywheel and clutch assembly.

If you have a crank that has suffered a spun rod bearing, damaged main bearing, or severe crank walk, recycle it and find another one. For many years, it was standard practice to resize crankshaft journals from engines that had suffered a spun bearing or other damage to one of the journals. Many rebuilders turn the crank as a matter of course to get a new, smooth journal surface for the bearing to ride on. However, this is not a good idea. Mitsubishi specifies crankshaft replacement when one of the journals wears, and this is not just to increase parts sales. There are a couple of sound mechanical reasons.

First, the stock 6-bolt and Evo cranks are chemically case hardened. The process is called nitriding, and involves heating the crankshaft to more than 900 degrees Fahrenheit and surrounding it with an ammonia atmosphere. The resulting hardening goes roughly .020 inch into the surface of the metal. What it does not do is make the crank any stronger. A nitrided crankshaft is just as strong as a non-nitrided crankshaft, but the surface is harder and more resistant to wear.

Second, regrinding damages the critical fillet radii (the small radiused groove around the edges of each

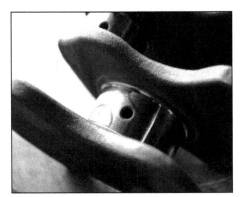

This is the kind of crankshaft damage you need to check for. This particular crank came from an engine that had water in the oil, and much of the pitting is from corrosion. If you're building a high-performance 4G63t, throw this crankshaft away and get another one.

Look closely at this stock, uncut crankshaft. The fillet radius around this journal is precisely cut and consistent all the way around. The stock Mitsubishi fillet radius is on the order of 3 mm, or about .120 inch.

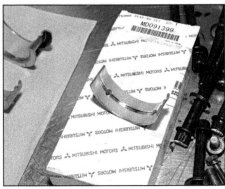

The OEM engine bearings are good quality parts that can be used for a high-performance rebuild. While 7-bolt bearings are supplied in small steps to selectively fit to each bearing cap or connecting rod and crankshaft combination to get the right clearance, replacement 6-bolt bearings are sold in increments of .25 mm only.

crank journal). A crankshaft that's cut too deep will loose the fillet radius, which weakens the crankshaft significantly. It requires a good crankshaft grinder to make a perfect fillet radius on a recut crank. That said, if you're building a street motor it is perfectly acceptable to use a reground crankshaft if you don't have a good alternative.

Aftermarket Crankshafts

The best crankshaft for most 2.0-liter rebuilds is a clean stock 88-mm Mitsubishi forging. Aftermarket crankshafts are nice, since they are sometimes lighter than the stock crank, but most engines will be fine with a stocker. The stock 4G63t crank is one of the strongest OEM forgings out there. There's no reason to use an aftermarket crank unless you're going endurance racing or can't find a good used stock crank.

The situation is different for stroker motors. The stock 100-mm crankshaft is not as strong as the 88-mm crankshaft because the rod and main bearing journals do not overlap as much. However, they have

proven to be reliable in 2.5- and 2.4-liter engines making as much as 500 to 600 hp.

Aftermarket stroker crankshafts with 100-mm and even 102-mm stroke are available. These are generally stronger than the stock Mitsubishi crankshaft, but they can be expensive. Look for a good quality forging from a reputable company like Brian Crower, TODA, or Cosworth.

Engine Bearings

Stock engine bearings are excellent quality, but aftermarket bearings are also available. Clevite 77 bearings are well-known American bearings, but Japanese bearings like ACL and OEM Mitsubishi are finished to a higher standard.

Choose one that you like and don't worry too much about it. Make sure that your bearings match the crankshaft and rods, as some aftermarket rods require the "wrong" bearings. For example, Eagle rods for the 7-bolt engine are designed for the 6-bolt bearings.

Aftermarket bearings are not normally selective fit; they only come one size. In practice, this is not a problem. The factory specification for bearing clearance is .0008 to .002 inch. Aim for the looser side (.002 inch) on a performance rebuild and accept that the bearings may not last as long as they would at a tighter clearance (which can destroy bearings under hard use).

Buy a tube of quality assembly lube (we prefer Redline) and don't be afraid to use it on and between every moving part. Coat bearings and shafts with it during assembly.

Install bearings carefully into clean, lint-free bearing housings. Bearings depend on a consistent, strong crush load to stay in place, and the slightest bit of dirt or lint can cause the bearing to pinch and spin on the shaft.

Check for crankshaft thrust clearance with a dial indicator. The factory specification is .002–.0098 inch of clearance. We prefer between .005 and .006 inch for optimized bearing life. Thrust clearance can be increased by removing material from the thrust bearing.

Use Plastigage to check bearing clearances; it is very accurate and easier to use than an internal micrometer or bore gauge. Don't forget to install the bearings in the proper orientation. Upper main bearings have an oil groove, and lower bearings don't, except for the center main, which has grooves all the way around.

Crank Walk

No discussion of 4G63t crankshafts would be complete without a thorough discussion of thrust bearing problems. Early 7-bolt motors have gained a reputation for having excessive crankshaft thrust bearing wear. This problem is commonly referred to as crank walk because of the way the crankshaft floats or walks back and forth in the block.

The first symptoms are often an inconsistent or variable clutch engagement point, or a clicking noise from the front of the engine. The symptoms usually get worse, until the crank moves out far enough to destroy the CPS under the timing cover. That stops the engine and the car has to be towed.

The reasons for crank walk have been debated since the first reported failure. Most reports of crank thrust failure, or crank walk, are of 2g manual transmission cars, but there have been reported instances of 1g 7-bolt

Use a feeler gauge to make sure the rod side clearance isn't excessive. If the clearance is too tight, the rods can be narrowed. If it's too loose, you must swap parts around. You're looking for between .004 and .010 inch of side clearance. We prefer clearance closer to .010 inch.

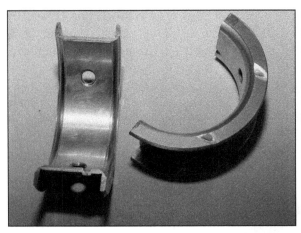

One of the most interesting things about crank walk is that it is almost exclusive to engines built before 1998. In 1998, Mitsubishi changed the thrust bearing to a 4-piece design from the original 2-piece (shown here).

engines failing, as well as automatics. The hysteria around crank walk reached such a point that many DSMers won't consider using a 7-bolt engine for a performance application because of the possibility of this problem occurring.

The truth is, it is a common problem for many different engines, and it's likely that the 2g 7-bolt 4G63t is only a little more prone to crank walk than other engines. The Internet has a tendency to amplify bad news, since people who do not suffer crankshaft problems aren't likely to post that their engines are fine.

People have gone to great lengths to determine why this happens, even to the point of cutting apart engine blocks (Magnus Motorsports) and removing the oil squirters. Mitsubishi changed many details of

the 4G63t engine when it designed the 7-bolt motor, so it could be almost anything. Some of the theories that have come up over the years include plugged oil-squirter check valves that reduce oil flow to the main bearings and thrust bearing, poorly designed thrust bearings that don't carry oil well enough, badly machined blocks with misaligned main bearings, badly machined crankshafts or poor crankshaft heat treating, bad clutch adjustment, and heavy aftermarket clutch pressure plates.

Crank walk can usually be avoided or eliminated during engine rebuilds by careful assembly and preparation. The first step is to use a perfect, unground crankshaft. Prepare the crankshaft by polishing and nitriding the surface, if it's within your budget. This will help it with-

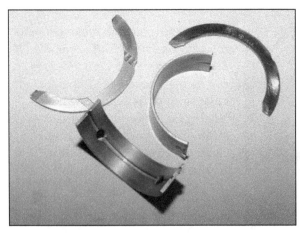

The 4-piece bearing assembly allows the thrust bearings to "float" and self-align with the machined thrust surface on the crankshaft. These bearings carried over to the Evo IV and newer 4G63t (and 4G64) engines too, and these later engines have very few thrust bearing failures.

stand dirt and wear much better.

If you have an early 7-bolt engine, use the best bearings you can find and carefully align the main bearing saddle with the thrust surface of the crankshaft. Do not use a crankshaft with any scoring or marks on the thrust surface. Use a new oil pump on any rebuild to ensure maximum oil pressure at idle, and use a large, quality oil filter to keep junk out of the oil passages in the block.

Minimize external contributing factors to crank walk by not using a heavy pressure plate. Use the lightest one you can get away with, with a "soft" clutch disc for maximum holding power. Make sure that your clutch is properly adjusted—there should be no pressure on the clutch release fork when the pedal is not depressed. Verify that this is the case even when the car is hot and you've been using the clutch all day. As the clutch fluid heats up, it expands and can cause the clutch hydraulic system to put slight pressure on the clutch. This is enough to cause excessive wear on the thrust bearing.

Finally, disconnect the clutch starter interlock. When the engine is first started, there is no oil on the thrust bearing surface, and if the clutch is pressed in when this happens, it wears the thrust bearing quickly. Disconnect the switch on the clutch pedal that forces you to hold the clutch in when starting, and always start the car in neutral. Don't press in the clutch until it's running and pressure registers on the gauge.

If you take all these steps, you're very unlikely to see a thrust bearing failure. Of course it should go without saying by now that you should pay attention to all the right clearances when installing the crankshaft and bearings—good assembly processes will help to get rid of crank walk.

SOURCE GUIDE

Advanced Clutch Technology
Clutches and flywheels
206 East Avenue K-4
Lancaster, CA 93535
www.advancedclutch.com

Advanced Engine Management (AEM)
Engine management systems, intake
manifolds, and fuel system parts
2205 126th Street, Unit A
Hawthorne Ca. 90250AEM
www.aempower.com

Blouch Turbo
Turbos and turbo kits
1925 State Route 72 North
Lebanon, PA 17046
www.blouchturbo.com

Brian Crower Inc.
Internal engine parts for 4G63 engines
P.O. Box 19066
San Diego, CA 92159
www.briancrower.com

Cometic Gasket
Custom and off-the-shelf gaskets
including MLS headgaskets for 4G63t
applications
8090 Auburn Road
Concord, OH 44077
www.cometic.com

Cosworth
Performance engine parts, including Evo
VIII and IX 4G63t
3031 Fujita Street
Torrance, CA 90505
www.cosworthusa.com

DEJON Tool
Intercooler plumbing and parts
8750 Covington-Bradford Rd
Covington, Ohio 45318
www.dejonpowerhouse.com

Eagle Specialty Products
Connecting rods and stroker cranks
8530 Aaron Ln
Southaven, MS 38671
www.eaglerod.com

ECMTuning
Manufacturer of DSMLink
1517 W Patrick St B13
Frederick, MD 21702
www.dsmlink.com

Forced performance
Turbos and turbo kits
671 New Hope Road West
McKinney, TX 75071
.forcedperformance.net

Full Throttle Speed
Manufacturer of MAF translator
34600 Klein Rd.
Fraser, MI 48026
http://www.fullthrottlespeed.com

GReddy Performance Products
Intake and exhaust parts
9 Vanderbilt
Irvine CA 92618
www.greddy.com

HKS USA
Cams, turbos and other performance
parts for DSM and Evo
13401 S. Main Street
Los Angeles, CA 90061
www.hksusa.com

JE Pistons
Custom and off-the-shelf pistons and rings
15312 Connector Lane
Huntington Beach, CA 92649
www.jepistons.com

Magnus Motorsports
Intake manifolds, DSM and Evo parts,
and tuning
8600 Keele St., Unit # 33
Concord, Ontario
Canada L4K 4H8
www.magnusmotorsports.com

Mitsubishi Motors North America
(MMNA)
Oversees US operations of Mitsubishi
Motors Corporation
6400 Katella Ave.
Cypress, CA 90630
mitsubishicars.com

Precision Turbo
Turbos and turbo kits
616A South Main Street
Hebron, IN 46341
www.precisionturbo.net

RC Engineering
Fuel Injectors and Fuel Injector Servicing
20807 Higgins Court
Torrance, CA 90501
www.rceng.com

Road/Race Engineering
DSM and Evo performance parts,
service, engine building, and race prep
13022 La Dana Ct.
Santa Fe Springs, Ca. 90670
www.roadraceengineering.com

Tactrix
Reflash and OBD interface cables and
software for Lancer Evolution
337 17th Avenue East
Seattle, WA 98112-5106
www.tactrix.com

TODA Racing USA
Internal engine parts for 4G63 engines,
including stroker kits, cams, and
valvetrain
18410 Bandilier Circle
Fountain Valley, CA 92708
www.todaracing.com

Wiseco pistons
7201 Industrial Park Blvd.
Mentor, OH 44060-5396
www.wiseco.com

Zeitronix
Wideband O_2 sensors and datalogging
equipment
P.O. Box 9125
San Pedro, CA 90734
www.zeitronix.com

MORE GREAT TITLES AVAILABLE FROM CARTECH®

CHEVROLET

How To Rebuild the Small-Block Chevrolet* (SA26)
Chevrolet Small-Block Parts Interchange Manual (SA55)
How To Build Max Perf Chevy Small-Blocks on a Budget (SA57)
How To Build High-Perf Chevy LS1/LS6 Engines (SA86)
How To Build Big-Inch Chevy Small-Blocks (SA87)
How to Build High-Performance Chevy Small-Block Cams/Valvetrains (SA105)
Rebuilding the Small-Block Chevy: Step-by-Step Videobook (SA116)
High-Performance Chevy Small-Block Cylinder Heads (SA125P)
High Performance C5 Corvette Builder's Guide (SA127)
How to Rebuild the Big-Block Chevrolet* (SA142P)
How to Build Max-Performance Chevy Big Block on a Budget (SA198)
How to Restore Your Camaro 1967–1969 (SA178)
How to Build Killer Big-Block Chevy Engines (SA190)
How to Build Max-Performance Chevy LT1/LT4 Engines (SA206)
Small-Block Chevy Performance: 1955-1996 (SA110P)
How to Build Small-Block Chevy Circle-Track Racing Engines (SA121P)
High-Performance C5 Corvette Builder's Guide (SA127P)
Chevrolet Big Block Parts Interchange Manual (SA31P)
Chevy TPI Fuel Injection Swapper's Guide (SA53P)

FORD

High-Performance Ford Engine Parts Interchange (SA56)
How To Build Max Performance Ford V-8s on a Budget (SA69)
How To Build Max Perf 4.6 Liter Ford Engines (SA82)
How To Build Big-Inch Ford Small-Blocks (SA85)
How to Rebuild the Small-Block Ford* (SA102)
How to Rebuild Big-Block Ford Engines* (SA162)
Full-Size Fords 1955–1970 (SA176)
How to Build Max-Performance Ford FE Engines (SA183)
How to Restore Your Mustang 1964 1/2–1973 (SA165)
How to Build Ford RestoMod Street Machines (SA101P)
Building 4.6/5.4L Ford Horsepower on the Dyno (SA115P)
How to Rebuild 4.6/5.4-Liter Ford Engines (SA155P)
Building High-Performance Fox-Body Mustangs on a Budget (SA75P)
How to Build Supercharged & Turbocharged Small-Block Fords (SA95P)

GENERAL MOTORS

GM Automatic Overdrive Transmission Builder's and Swapper's Guide (SA140)
How to Rebuild GM LS-Series Engines* (SA147)
How to Swap GM LS-Series Engines Into Almost Anything (SA156)
How to Supercharge & Turbocharge GM LS-Series Engines (SA180)
How to Build Big-Inch GM LS-Series Engines (SA203)
How to Rebuild & Modify GM Turbo 400 Transmissions (SA186)
How to Build GM Pro-Touring Street Machines (SA81P)

MOPAR

How to Rebuild the Big-Block Mopar (SA197)
How to Rebuild the Small-Block Mopar* (SA143P)
How to Build Max-Performance Hemi Engines (SA164)
How to Build Max-Performance Mopar Big Blocks (SA171)
Mopar B-Body Performance Upgrades 1962-1979 (SA191)
How to Build Big-Inch Mopar Small-Blocks (SA104P)
High-Performance New Hemi Builder's Guide 2003-Present (SA132P)

OLDSMOBILE/ PONTIAC/ BUICK

How to Build Max-Performance Oldsmobile V-8s (SA172)
How To Build Max-Perf Pontiac V8s SA78
How to Rebuild Pontiac V-8s* (SA200)
How to Build Max-Performance Buick Engines (SA146P)

SPORT COMPACTS

Honda Engine Swaps (SA93)
Building Honda K-Series Engine Performance (SA134)
High-Performance Subaru Builder's Guide (SA141)
How to Build Max-Performance Mitsubishi 4G63t Engines (SA148)
How to Rebuild Honda B-Series Engines* (SA154)
The New Mini Performance Handbook (SA182P)
High Performance Dodge Neon Builder's Handbook (SA100P)
High-Performance Honda Builder's Handbook Volume 1 (SA49P)

*Workbench® Series books featuring step-by-step instruction with hundreds of color photos for stock rebuilds and automotive repair.

ENGINE

Engine Blueprinting (SA21)
Automotive Diagnostic Systems: Understanding OBD-I & OBD II (SA174)

INDUCTION & IGNITION

Super Tuning & Modifying Holley Carburetors (SA08)
Street Supercharging, A Complete Guide to (SA17)
How To Build High-Performance Ignition Systems (SA79)
How to Build and Modify Rochester Quadrajet Carburetors (SA113)
Turbo: Real World High-Performance Turbocharger Systems (SA123)
How to Rebuild & Modify Carter/Edelbrock Carbs (SA130)
Engine Management: Advanced Tuning (SA135)
Designing & Tuning High-Performance Fuel Injection Systems (SA161)
Demon Carburetion (SA68P)

DRIVING

How to Drift: The Art of Oversteer (SA118P)
How to Drag Race (SA136)
How to Autocross (SA158P)
How to Hook and Launch (SA195)

HIGH-PERFORMANCE & RESTORATION HOW-TO

How To Install and Tune Nitrous Oxide Systems (SA194)
Custom Painting (SA10)
David Vizard's How to Build Horsepower (SA24)
How to Rebuild & Modify High-Performance Manual Transmissions* (SA103)
High-Performance Jeep Cherokee XJ Builder's Guide 1984–2001 (SA109)
How to Paint Your Car on a Budget (SA117)
High Performance Brake Systems (SA126P)
High Performance Diesel Builder's Guide (SA129)
4x4 Suspension Handbook (SA137)
How to Rebuild Any Automotive Engine* (SA151)
Automotive Welding: A Practical Guide* (SA159)
Automotive Wiring and Electrical Systems* (SA160)
Design & Install In Car Entertainment Systems (SA163)
Automotive Bodywork & Rust Repair* (SA166)
High-Performance Differentials, Axles, & Drivelines (SA170)
How to Make Your Muscle Car Handle (SA175)
Rebuilding Any Automotive Engine: Step-by-Step Videobook (SA179)
Builder's Guide to Hot Rod Chassis & Suspension (SA185)
How To Rebuild & Modify GM Turbo 400 Transmissions* (SA186)
How to Build Altered Wheelbase Cars (SA189)
How to Build Period Correct Hot Rods (SA192)
Automotive Sheet Metal Forming & Fabrication (SA196)
Performance Automotive Engine Math (SA204)
How to Design, Build & Equip Your Automotive Workshop on a Budget (SA207)
Automotive Electrical Performance Projects (SA209)
How to Port Cylinder Heads (SA215)
Muscle Car Interior Restoration Guide (SA167)
High Performance Jeep Wrangler TJ Builder's Guide: 1997-2006 (SA120P)
Dyno Testing & Tuning (SA138P)
How to Rebuild Any Automotive Engine (SA151P)
Muscle Car Interior Restoration Guide (SA167P)
How to Build Horsepower - Volume 2 (SA52P)
Bolt-Together Street Rods (SA72P)

HISTORIES & PERSONALITIES

Fuelies: Fuel Injected Corvettes 1957–1965 (CT452)
Yenko (CT485)
Lost Hot Rods (CT487)
Grumpy's Toys (CT489)
Rusted Muscle — A collection of junkyard muscle cars. (CT492)
America's Coolest Station Wagons (CT493)
Super Stock — A paperback version of a classic best seller. (CT495)
Rusty Pickups: American Workhorses Put to Pasture (CT496)
Jerry Heasley's Rare Finds — Great collection of Heasley's best finds. (CT497)
Street Sleepers: The Art of the Deceptively Fast Car (CT498)
Ed 'Big Daddy' Roth — Paperback reprint of a classic best seller. (CT500)
Rat Rods: Rodding's Imperfect Stepchildren (CT486)
East vs. West: Rods, Customs Rails (CT501)
Car Spy: Secret Cars Exposed by the Industry's Most Notorious Photographer CT502)

CarTech®, Inc. 39966 Grand Ave., North Branch, MN 55056. Ph: 800-551-4754 or 651-277-1200 • Fax: 651-277-1203
Brooklands Books Ltd., PO Box 146 Cobham, Surrey KT11 1LG, England. Ph: 01932 865051 • Fax 01932 868803
Brooklands Books Aus., 3/37-39 Green Street, Banksmeadow, NSW 2019, Australia. Ph: 2 9695 7055 • Fax 2 9695 7355

Visit us online at
www.cartechbooks.com for more info!

Additional books that may interest you...

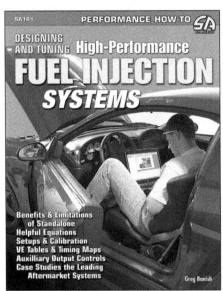

DESIGNING AND TUNING HIGH-PERFORMANCE FUEL INJECTION SYSTEMS *by Greg Banish* Engineer and industry veteran Greg Banish, author of the best-selling title *Engine Management*, tackles this complex subject and explains it in an easy-to-read manner. Learn useful formulas, VE equation and airflow estimation, and more. Also covered are setups and calibration, creating VE tables, creating timing maps, auxiliary output controls, start-to-finish calibration examples with screen shots to document the process. Useful appendixes include a glossary and a resource guide. Aftermarket standalone systems are a great way to dial in performance and reliability. This is the book you need to become an expert in this popular modification. Softbound, 8.5 x 11 inches, 128 pages, 250 color photos. *Item # SA161*

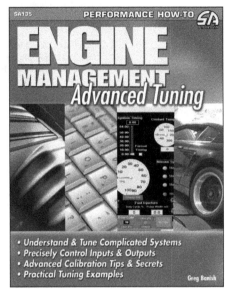

ENGINE MANAGEMENT: ADVANCED TUNING *by Greg Banish* As tools for tuning modern engines have become more powerful and sophisticated in recent years, the need for in-depth knowledge of engine management systems and tuning techniques has grown. This book takes engine-tuning techniques to the next level, explaining how the EFI system determines engine operation and how the calibrator can change the controlling parameters to optimize actual engine performance. It is the most advanced book on the market, a must-have for tuners and calibrators and a valuable resource for anyone who wants to make horsepower with a fuel-injected, electronically controlled engine. Softbound, 8.5 x 11 inches, 128 pages, 250 color photos. *Item # SA135*

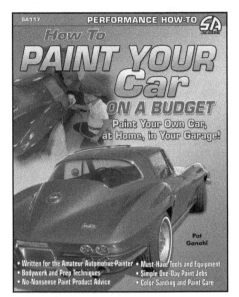

HOW TO PAINT YOUR CAR ON A BUDGET *by Pat Ganahl* If your car needs new paint, or even just a touch-up, the cost involved in getting a professional job can be more than you bargained for. Author Pat Ganahl unveils dozens of cost-saving secrets that help you paint your own car. From simple scuff-and-squirt jobs to full-on, door-jambs-and-all paint jobs, Ganahl covers everything you need to know to get a great-looking coat of paint on your car and save lots of money in the process. Covers painting equipment, the ins and outs of prep, masking, painting and sanding products and techniques, and real-world advice on how to budget wisely when painting your own car. Softbound, 8.5 x 11 inches, 128 pages, 400 color photos. *Item # SA117*

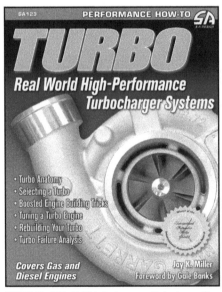

TURBO: Real-World High-Performance Turbocharger Systems *by Jay K. Miller* Gas or Diesel, 4, 6, 8, or more cylinders, this book shows you how to test, install, and maintain your high-performance turbo system. Learn how turbochargers work, how to choose the right turbo or turbos for your engine by reading flow maps, and how to tune your engine to run perfectly with your turbo system. Author Jay K. Miller discusses the various components of a turbocharger and explains how to decode complicated turbocharger model numbers, compressor maps, and other specs, and if you run into problems with your turbo system, there's also a detailed chapter on failure analysis to help you figure out what's wrong and how to fix it. Softbound, 8.5 x 11 inches, 160 pages, 320 color photos. *Item # SA123*

CHECK OUT CARTECH'S NEW, IMPROVED WEB SITE!

- Find helpful tech tips & articles
- Get bonus material from our books
- Browse expanded e-book selection
- Join the discussion on our blog
- Look inside books before you buy
- Rate & review your CarTech collection
- Easy, user-friendly navigation
- Sign up to get e-mails with special offers
- Lightning fast, spot-on search results
- Secure online ordering
- 24/7 access

- Check out our Featured Weekly Ride
- Reader's Rides – Show off your car!

www.cartechbooks.com

www.cartechbooks.com or 1-800-551-4754